Green Outcomes in the Real World

Green Outcomes in the Real World

Global Forces, Local Circumstances, and Sustainable Solutions

PETER MCMANNERS

GOWER

Gower Applied Business Research
Our programme provides leaders, practitioners, scholars and researchers with thought provoking, cutting edge books that combine conceptual insights, interdisciplinary rigour and practical relevance in key areas of business and management.

Published by
Gower Publishing Limited
Wey Court East
Union Road
Farnham
Surrey, GU9 7PT
England

Ashgate Publishing Company
Suite 420
101 Cherry Street
Burlington,
VT 05401-4405
USA

www.gowerpublishing.com

British Library Cataloguing in Publication Data
McManners, Peter J.
Green outcomes in the real world : global forces, local circumstances, and sustainable solutions.
1. Globalization--Environmental aspects. 2. Environmental policy. 3. Environmental economics. 4. Sustainable development--International cooperation.
I. Title
363.7'0526-dc22

ISBN: 978-0-566-09179-7 (hbk)
978-0-566-09180-3 (ebk)

Library of Congress Control Number: 2010935403

Printed and bound in Great Britain by
MPG Books Group, UK

Contents

List of Figures, Tables and Boxes		*vii*
List of Abbreviations		*ix*
Reviews for Green Outcomes in the Real World		*xiii*
Foreword		*xv*
Preface		*xvii*
PART 1	THE PRINCIPLES OF A GREEN FUTURE	
1	**Sustainability in a Globalized World**	**3**
2	**Sustainability**	**21**
3	**Subsidiarity**	**35**
4	**The Primacy of the State**	**47**
5	**Green Economics**	**55**
PART 2	CHANGING SOCIETY AND THE ECONOMY	
6	**Return to a Natural World Order**	**75**
7	**Global Commodity Flows**	**83**
8	**Population Dynamics**	**99**
9	**Finance and Capital Flows**	**111**
10	**The Global Knowledge Economy**	**131**
PART 3	A CHANGING WORLD	
11	**Global Cooperation and Coordination**	**143**

12 **The Key Role of Government** **157**

13 **Global Corporations** **169**

14 **Human-Scale Communities** **181**

15 **Sustainable By Design** **195**

Epilogue *209*
References *213*
Index *223*

List of Figures, Tables & Boxes

Figure 2.1	The Two-Stool Model of Sustainability	29
Table 8.1	Ecological deficits and surpluses	105
Box 7.1	The Milk Road Map 2008	91
Box 12.1	Breakthroughs for the Twenty-First Century	164
Box 13.1	The Dutch East India Company	174

List of Abbreviations

ARRF	Acid Rain Retirement Fund
ASEAN	Association of Southeast Asian Nations
BedZED	Beddington Zero Energy Development
CDM	Clean Development Mechanism
CDO	Collateralized Debt Obligation
CSR	Corporate Social Responsibility
CTE	Committee on Trade and Environment
DECC	Department of Energy and Climate Change
DEFRA	Department for Environment, Food and Rural Affairs
DTIE	Division of Technology, Industry and Economics
EPA	Environmental Protection Agency
EPI	Environmental Performance Index
ERM	European Exchange Rate Mechanism
ESI	Environmental Sustainability Index
EU	European Union
EU ETS	European Union Emission Trading System
FDI	Foreign Direct Investment

FSB	Financial Stability Board
FSF	Financial Stability Forum
GATT	General Agreement on Tariffs and Trade
GDP	Gross Domestic Product
GGND	Global Green New Deal
gha	global hectares
GM	General Motors
GNH	Gross National Happiness
GNP	Gross National Product
HDI	Human Development Index
HIPC	Heavily Indebted Poor Countries
HPI	Happy Planet Index
IFG	International Forum on Globalization
IMF	International Monetary Fund
IPCC	Intergovernmental Panel on Climate Change
IPR	intellectual property rights
LETS	Local Exchange Trading Schemes
MBA	Master of Business Administration
MDG	Millennium Development Goals
MEA	multilateral environmental agreement

MNC	multinational corporation
NAFTA	North American Free Trade Agreement
NDPB	non-departmental public body
NEF	New Economics Foundation
NGO	non-governmental organization
NOAA	National Oceanic and Atmospheric Administration
OECD	Organisation for Economic Co-operation and Development
PETM	Paleocene-Eocene Thermal Maximum
plc	public limited company
PPP	purchasing power parity
PV	photovoltaic
SDC	Sustainable Development Commission
SDR	Special Drawing Right
SDRT	Stamp Duty Reserve Tax
SSM	Special Safeguard Mechanism
UN	United Nations
UNCTAD	United Nations Conference on Trade and Development
UNDP	United Nations Development Programme
UNEP	United Nations Environment Programme
VAT	Value Added Tax

VOC	Vereenigde Oost-indische Compagnie
WCED	World Commission on Environment and Development
WEEE	Waste Electrical and Electronic Equipment
WMO	World Meteorological Organization
WRI	World Resources Institute
WSSD	World Summit on Sustainable Development
WSTO	World Sustainable Trade Organization
WTO	World Trade Organization

Reviews for Green Outcomes in the Real World

'In this ground-breaking book, Peter McManners shines a spotlight on some of the most intractable and important issues of the current age. He exposes the "dark underbelly" of globalization and the problems that arise from the narrow pursuit of economic objectives. He is not alone in recognising the problems, but the strength of this book is the policy framework proposed to solve them.

In examining the world economy and society, McManners looks for ways to reduce the pressure on the environment through solutions that improve human society. Where he writes about economics, the discussion takes place within the broader context of a sustainable human society. Few writers could match the breadth of perspective that McManners brings to this important debate.

McManners goes beyond the confines of accepted knowledge, not through adding complexity to existing theory, but by returning to the basic constructs of how to organise human affairs. In doing so he avoids jargon. The book is accessible to a wide audience. The result is a firm foundation on which to build an economy and society fit for the challenges of the twenty-first century. The core argument, that the prime focus of government should be to craft policy to implement social and environmental outcomes, is hard to dispute.

Green Outcomes in the Real World *is published at a time when the world needs new ideas and a new direction. It defines that new direction and should be compulsory reading for policy makers and students of international affairs. It sets the framework that could support the greening of the economy and society.'*

Paul Taylor, Emeritus Professor, London School of Economics

'This excellent book provides a unique perspective on sustainability and globalization, two subjects which are usually viewed as inherently contradictory. The author argues convincingly that the two can be reconciled if we are willing to introduce some fundamental changes in the way we run our economy and society. He provides some powerful and bold insights into how to achieve this.'

Professor Emilio Herbolzheimer, Henley Business School,
Reading University

'Peter McManners rightly identifies the failure to question the orthodoxy of economic globalisation as a roadblock along humanity's journey towards sustainability.

Committed free-traders will be uncomfortable with McManners' conclusions, but they should read this book. They will find he is no enemy of competition and free markets. The essence of his valuable proposal is that by encouraging a greater diversity of economic models and allowing more room for nations individually, and in self-selecting groups, to find their own paths to sustainability, competition will encourage creativity, innovation and a more resilient global economy.

McManners outlines a new global economic structure, built on the foundations of his four principles of proximization, which not only sits more easily within its surrounding eco-systems, but is also a more comfortable and pleasing place for us to live.'

Stewart Wallis, Executive Director, the New Economics Foundation

'In this valuable and incisive book, McManners provides a coherent overview of thinking about the critical issue of sustainability in the real world. He underlines the need to adapt economics to serve society better and to reorient the paths of globalisation and growth so as to decouple our notion of progress from environmental degradation. To achieve such deep changes, he makes challenging proposals for a new balance of responsibility and partnership in world affairs. These are indeed critical issues which will determine the future of humanity in the twenty-first century.'

Martin Lees, Secretary General, Club of Rome

Foreword

Green Outcomes in the Real World brings the concept of a green economy in from the periphery of policy formulation to take centre stage. The result is surprising in that McManners dismantles the edifice of economic globalization, and satisfying, in that the result shows a way forward to solve the most difficult dilemma of our age: reconciling economic aspirations with the imperative to safeguard the environment.

The buzzword of political economy has long been globalization. Neo-liberals have jumped onto this band wagon with their ideology of free market economics. American neo-conservatives have used it to finesse a grab for increased world power for the United States and American corporations. Globalization, they claim, will make us richer, or more successful, or, even, greener.

But there have also long been those who have seen its problems. The wealth has not been shared. There has been very little trickle down and a lot of trickle up. Two pressing current crises have raised further doubts: the desperate sickness of the international financial system after 2008, and the problems that threaten environmental catastrophe, especially the increases in carbon emissions and global warming, with all the tragic consequences that could follow. This book concentrates on the latter, arguing, convincingly, that economic theory has to bend to the higher aim of safeguarding the ecosystem now and for future generations.

McManners ponders ways to achieve a greener planet in which resources are husbanded, environmental problems are dealt with and the interests of individuals are stressed ahead of increasing Gross Domestic Product. He considers an alternative approach to globalization. Solutions can be pursued from the bottom up rather than the top down, and problems are more likely to be solved among neighbours at the local level, for good practical reasons of mutual advantage, rather than through universal rules and principles drawn up in remote international organizations and grand conferences.

McManners calls this a policy of *proximization*. It should be introduced into as wide a range of areas as possible, including trade policy, the supply of food and security in the broadest sense, as well as the more direct limitation of the ecological capacity. The primary attention to the world's problems should be in areas where success is more likely, in viable nation states, or in regions such as the European Union. The relevant principle is that introduced into the European Union in the Treaty of Maastricht, but now to be applied at the global level: the principle of *subsidiarity*.

Leading by example, following the framework of proximization, is likely to be emulated in other places when the results are plain to see. The example that the developed world has set in pursuing the policies of economic globalization has to change. This book should be read and discussed widely to invigorate the debate about the future direction for the economy and society.

McManners has written a valuable critique of the easy assumptions of fundamentalist globalization, especially as they are applied to environmental problems. The arguments are appealing as a practical agenda. If adopted his approach may lead to a less integrated global system, but this is not the point. The world could be more integrated at the local level, where problems are more easily managed and *sufficiently* integrated at the global level. What is needed is a more civilized agenda, a sustainable management of the environment, rather than the mindless pursuit of wealth for the few in a global free market.

This book is to be strongly recommended as a sensitive exploration of ways forward at this point of economic and environmental crisis. It looks beyond the values which at present dominate the economies and societies of the developed world, to a world in which there is a more secure future for everyone.

Paul Taylor,
Emeritus Professor of International Relations,
London School of Economics.

Preface

Modern civilization is the pinnacle of human achievement. Through the nineteenth and twentieth centuries human ingenuity built the greatest civilization of all time. Our technology is advancing at such a pace that it seems there is nothing we cannot achieve. One of the secrets of our success has been the development of economic theory to provide a sound basis for organizing society and allocating resources efficiently. The particular idea that has accelerated progress and brought such wealth and material improvement in human lives over the last three decades is economic globalization, but the time has come to consider a new direction for society.

Humans are quick-witted, dexterous and can work well together in groups. This is why the species Homo sapiens is so successful. However, we can also be blind to problems until they grow large enough to grab our attention. We have been slow to understand that our narrow focus on economic outcomes is undermining the integrity of the biosphere. Now that this is understood, we have to step back and consider a policy framework to replace economic globalization. This need not be a step back in time to revert to a past age. It should be a metaphorical step back to regain our perspective on priorities for society. From this new vantage point, we can clearly see the dangers and plan the leap forward that is required to move beyond today's problems to the next phase of human progress.

I occupy a no-man's land between proponents of economic globalization and those who campaign for protection of the global environment. I can see clearly both sides of the debate, but the debate has diverged. It is imperative that we pull the debate together and reconcile our material aspirations with the pressing need to relieve the pressure on the natural systems of the planet.

Conventional economists tell us that economic globalization is good economics, and so it is. Globalization has brought economic success to many countries that have fully embraced the opportunities and accepted the harsh discipline of the open world market.

Environmentalists complain that the way modern globalized society operates is damaging the ecosystem; this also is true. Some of the damage we have caused is already irreversible, and there is no sign that humans are backing off. The evidence shows that the damage is accelerating.

I argue in this book that fundamental change is required in the way we run the economy and society. The policies of the last 20 years have, in many ways, been successful. They have shown the extraordinary power of laissez-faire capitalism to transform economies and lift people out of poverty, but we are deluded if we remain blind to the dark underbelly of associated problems that arise. Overcoming financial crisis ranks as a relatively minor challenge compared with the looming problems of climate change, oil depletion and population growth.

People want the triple-win of a strong economy, a healthy environment and a stable cohesive society. Over the first decade of the twenty-first century, I have heard linguistic contortions from world leaders and so-called experts attempting to justify the continuation of business as usual. Some seem to believe their own rhetoric that continued progress down the path of growth fuelled by economic globalization is the way to alleviate the problems of the world. This self-delusion is a convenient way to avoid facing the real challenges.

There is a simple truth that we ignore at our peril. Economic globalization brings short-term success but is diametrically opposed to the requirements of a long-term sustainable society. For the policies of economic globalization, there is a choice – to continue, or shift to a different paradigm. For sustainability there is no choice, or no sensible choice, if we believe that humanity has a future. Continuing to ignore the need to balance the economy and the environment will eventually destroy our species. This means that the new paradigm has to be based on the principle of sustainability. It has become clear that overlaying the concept of sustainability on the current economic system cannot work. When, instead, sustainability is inserted at the foundations, the resulting policy framework is radically different.

The aim is to set the context within which economists and policy makers can develop a new paradigm. I describe and explain a policy framework that encourages selfish determination to build sustainable societies. It appears to be a U-turn from the economic policies that have brought material advancement and wealth. In fact, it is a huge leap forward, providing the context within which humans can redirect their drive for success to build vibrant societies in tune with local resources and local customs.

As the inherent contradictions between globalization and sustainability are addressed and resolved, the old battle lines between the policies of globalization and protectionism will be blurred. For example, global cooperation and interconnections will be vital to support the migration of green technologies and processes across the planet. The term 'protectionism' will lose its negative connotations when it means safeguarding the environment and bolstering social cohesion within national economies.

Instead of fighting over divergent policy choices that separate the economy from the ecosystem, we should engage in building the future for human society and the planet. This reconciliation of global forces with local circumstances makes it possible to deliver a stable and sustainable world society.

My intention is to make my small contribution to ensuring that the world is a better place. Where I dismantle cherished theories, I do so because the world has changed and we need to adapt. Where I make mistakes, which is inevitable when arguing for radical change, I crave your indulgence and ask that you look beyond these to identify the alterations that will improve my framework.

In testing my ideas, I am indebted to Emeritus Professor Paul Turner, London School of Economics, and Professor Emilio Herbolzheimer, Henley Business School of Reading University, to be the first to read my manuscript. Their expert feedback has ensured that my green perspective remains firmly rooted in the challenges of the real world.

Whether you support or oppose my ideas, speak up and join the debate. This is an important and defining point in human history. What we do in the coming decades will have far-reaching consequences for the planet and for us all.

Peter McManners

PART 1

The Principles of a Green Future

1

Sustainability in a Globalized World

Economic globalization has transformed the world economy. At the same time, the global environment has come under increasing pressure. It is becoming increasingly difficult to balance economic aspirations with improvements in human welfare whilst at the same time conserving the integrity of the ecosystem.

As globalization matures and the immediate benefits work through the system, the downsides will become more apparent: the competition will become ever tougher, the fight over resources ever fiercer and the risks to the world environment ever greater. On the one hand, governments will urge their local industry to be more competitive in the global market; on the other, they will attempt to maintain social cohesion and counter threats to the environment.

When governments find that reconciling these conflicting demands is becoming increasingly difficult, many countries will lose their enthusiasm for globalization and seek to take closer control over managing their affairs and running their economies.

The policy framework that I present in this book reconciles the two big issues of our time: globalization[1] and sustainability. The former is the consequence of free trade and market liberalization. The latter is the aspiration to improve human quality of life concurrently with safeguarding the integrity of the Earth's ecosystem.

---

1 The term 'globalization' has other connotations such as the diffusion of culture or language, and has been used negatively against the spread of Western values or complaints about the intrusion of English words into other languages. To avoid this, in this book, 'globalization' is synonymous with 'economic globalization'.

Globalization has proved to work in delivering economic outcomes. Over the last two decades, the world economy has expanded on a massive scale as the policies of free trade and deregulated markets have been widely adopted. The sum total of world Gross Domestic Product (GDP)[2] more than doubled between 1987 and 2007, increasing to $65 trillion in 2007 (IMF 2008). Over the same period, world trade surged from $2.5 trillion to $14.0 trillion (WTO 2009a).

The Earth's ecosystem has come under increasing pressure concurrently with this economic expansion. The most visible and measurable consequence is climate change.

Climate Crisis

The climate crisis[3] has brought threats to the ecosystem higher up the political agenda. It has attracted a lot of discussion but little substantive, effective action. As the climate crisis grows in people's consciousness, it will be a trigger for starting a real dialogue over the nature of a sustainable world society. The Climategate furore that erupted in 2009 when 'mistakes' made by climate scientists were exposed has stifled the debate in 2010 but the facts have not changed.

Each year, levels of CO_2 rise to reach a new record. In 2008, the level was 386 ppm (Tans 2009); higher than at any time over the last 800,000 years (Lüthi et al. 2008). These elevated levels of CO_2 are the prime cause of climate change leading to rising sea levels (IPPC 2007). McGranahan et al. (2007) report that 10 per cent of the world's population live in low-lying coastal areas that are at risk of flooding. This implies that world society will have to make plans to find new homes for up to 600 million[4] people if robust evidence emerges of a significant rise in sea level over the coming decades. This will be an immense challenge for policy makers, but climate change could be far worse than the sober predictions of the Intergovernmental Panel on Climate Change (IPCC).[5]

2 GDP based on purchasing power parity (PPP).
3 'In my view, climate change is the most severe problem that we are facing today – more serious even than the threat of terrorism' (King 2004).
4 10 per cent of a world population of 6 billion people.
5 IPCC was set up by the World Meteorological Organization (WMO) and the United Nations Environment Programme (UNEP) in 1988 to provide an objective source of information about climate change.

It is necessary to reach way back into geological time for circumstances that are similar to those today. The fossil record shows that about 55 million years ago an event called the Paleocene-Eocene Thermal Maximum (PETM) occurred. At that time, temperatures rose rapidly by as much as 10°C at high latitudes in the Arctic and Antarctic. Tropical oceans and deep-ocean waters warmed by between four and six degrees. These were accompanied by dramatic effects on plants and animals. Scientists think that the fast-changing climate was driven by a natural release of carbon-containing greenhouse gases comparable to what has been occurring with the release of CO_2 and other gases since the start of the Industrial Revolution.

We should be very worried. The relatively low levels of global warming being observed in these early stages of climate change may be enough to start the process of thawing frozen deposits of methane hydrates. These are present in vast quantities as frozen sediments in the deep oceans and some are associated with permafrost soils in the Arctic. Warming of the oceans and the Arctic region could release sufficient methane to initiate a positive feedback effect, releasing still more gas as ocean waters and permafrost regions continue to warm. Methane is a much more potent greenhouse gas than CO_2. The theory is that the process reaches a tipping point when it starts to accelerate rapidly, without any way to stop the process, until the world reaches a new, and much hotter, equilibrium.

The Paleocene-Eocene Thermal Maximum is proof that the Earth's system does have tipping points which, when exceeded, can quickly run ahead to a new and different equilibrium. Exactly why these past episodes have occurred is subject to much speculation. Scientists can use bubbles of air trapped in the deep Antarctic ice to investigate ancient climate history, but this takes them back only 800,000 years. This is not far enough. Scientists cannot prove that the world's climate is, or is not, reaching a tipping point. So fears that there may be a tipping point are not featured in reports from the IPCC. The point that should concern us most is that if we are pushing against a tipping point, and we find it, there will be nothing we can do to stop the runaway climate change that would follow. This risk should worry us more than any economic crisis – and climate change is part of a larger problem

The Millennium Ecosystem Assessment

In 1998, the Millennium Ecosystem Assessment was set up through collaboration between the World Resources Institute (WRI), United Nations Environment

Programme (UNEP), the World Bank and United Nations Development Programme (UNDP). The report *Millennium Ecosystem Assessment 2005* was published seven years later. It found that:

> *Over the past 50 years, humans have changed these ecosystems more rapidly and extensively than in any comparable period of time in human history, largely to meet rapidly growing demands for food, fresh water, timber, fiber, and fuel. This transformation of the planet has contributed to substantial net gains in human well-being and economic development. But not all regions and groups of people have benefited from this process – in fact, many have been harmed. Moreover, the full costs associated with these gains are only now becoming apparent.*

The *Millennium Ecosystem Assessment* is a well-balanced analysis. It recognizes that 'transformation of the planet has contributed to substantial net gains in human well-being and economic development.' This is progress we will be reluctant to put at risk, but the report carries the warning: 'the full costs associated with these gains are only now becoming apparent'.

The members of the board overseeing the Millennium Ecosystem Assessment were drawn from the United Nations (UN), World Bank and a number of non-governmental organizations (NGOs). The assessment has influenced policy in the areas represented by the board members, such as environmental and biodiversity policy. For the report to have real impact, it would need to influence wide areas of world policy from international development cooperation to trade and financial sectors (The Royal Society 2006). There has been little evidence of this.

Reconciling Globalization and Sustainability

There is a fundamental conflict between globalization and sustainability. The policy of globalization is sucking resources from the Earth at an increasing rate, driving activities to places with the lowest environmental and social standards. On the other hand, the policy of sustainability is seeking to persuade individuals, organizations and corporations to behave responsibly towards social issues and the environment. I am not alone in recognizing that global capitalism is in need of reform (Hart 2005, Porritt 2005). But I go further than many other commentators in my belief that our existing mechanisms of capitalism, as currently implemented, cannot handle these diverging forces.

The established order of how we run society and its institutions is set to be overthrown. *The Sustainable Revolution* (McManners 2008) will cut across every aspect of society from infrastructure, such as energy systems and transport, to policies for communities, agriculture and trade. Although we cannot know the results, we can try to predict how the revolution will develop and identify parameters that are likely to nudge its progress towards desirable outcomes. We should aim to hang on to the best of globalization whilst incorporating the principles of sustainability.

It will be tough to reconcile globalization and sustainability at world level. First, it requires good global governance to oversee world standards and policies. But national governments tend not to bow down to the authority of global institutions, except where it suits their interests. Second, it needs a concept of global citizenship that is established and deep-rooted, so that the world's population will accept constraints for the global good. But people do not behave as global citizens. If we expect action to be based on global governance and global citizenship, then we should not expect too much.

The key to understanding how to tame globalization and adopt sustainability is to look at humankind's common concerns. These are the resources we all share: the atmosphere, the oceans and the biodiversity of life, which knows no borders. The problems connected with our shared resources were defined more than two decades ago by The World Commission on Environment and Development (WCED) (1987), leading to the setting up of UNEP.

Under the auspices of the UNEP, the world is trying to sort out its environmental governance, and seeking to persuade all of us to behave as responsible residents of the world. But the sum total of our actions has been disappointing. Far from starting to solve the problems, we have presided over an acceleration of environmental degradation. This apparent inability to act – despite good intentions and the backing of the world's highest decision-making body – should be of deep concern. We should be pressing for action rather than forever restating the situation.

There is still time to sort out our pressing problems but, without the mechanisms to do so, we are like a rabbit caught in the glare of the headlights of a speeding truck. We can easily escape, but we are transfixed by the situation. Instead of leaping off to the side, we wait to see if the wheels run over us.

The Tragedy of the Commons

We can use an example to get a feel for the leap that we should make. Let us consider an area of common land coming under the sort of pressure that the world is now encountering. (The parallel with the whole Earth system is only partial, but the lesson we can draw provides a useful insight.) The land has been common for as long as anyone can remember, and a number of families have grazing rights. To date, the common has been large enough, with an unwritten balance emerging through the generations that allows the common to support the community.

Now economic development is threatening the common. Many holders of grazing rights have realized that their incomes can be greater, their families larger and their lives materially better by maintaining larger herds of animals. Some families have already increased their herds substantially. Others see the benefits and decide to follow suit. It is only at this stage that the more enlightened holders of grazing rights start to see the risks. The common had been running well within its capacity, but it is now starting to show signs of overgrazing.

Over a protracted period, discussions take place to seek to manage the common. Some agreements to limit damage are reached. For the main issue, that of overgrazing, a large group of rights holders agree to control numbers, but this agreement has little chance of success if other rights holders continue to increase their herds. The common becomes subject to an inevitable cycle of decline. If the cycle is not broken, the grazing will be stripped bare, animals will die and the returns to all will diminish. If we liken our world to the common in our example, this is the decision point that the world has reached.

We can accept the situation and milk it whilst we can – before the inevitable decline destroys the resources of the world – or find another way. This is the classic problem described by Garret Hardin (1968). Hardin's proposed solution was through 'relinquishing the freedom to breed'. This was, and still is, a controversial statement, but Hardin was correct to identify that change is needed and that we might find aspects of it unpalatable.

Breaking the Spiral of Decline

In the example of the common, I see a way out, but it requires a break with tradition. A group of grazing-rights holders decide to ensure their own destiny by breaking the spiral of decline. They take control of part of the common and limit grazing and numbers in that area. This group of people need to robustly defend their stance and keep others off 'their' area. They hope that, over time, the evident success of better management of their part of the common will be replicated. They will need to work with and negotiate with the other rights holders to diffuse conflict but – as far as their influence and power extends – they must also be resolute in not allowing the cycle of decline to continue.

World society will respond to environmental problems in much the same way as the common rights holders. Sustainable societies will be built on a selfish determination to deliver a better life for a particular society's own members, moving towards a situation where economics, environmental protection and social provision are balanced. However, we should not underestimate the complexity of the challenge – and we should not expect that real-world solutions lead to Utopia.

Breaking the spiral of decline means facing up to some hard truths. This is not a course of action that most advocates of green policy like but there are issues that can no longer be ignored. At the heart of this tricky debate is the balance between economics, environmental protection and social provision.

Balancing Economics, Environmental Protection and Social Provision

These three legs of sustainability are interrelated (Chapter 2). We are beginning to understand that maximizing economic performance puts pressure on the environment and can undermine social structures. It is equally true that a narrow focus on social targets, such as the elimination of poverty, can put the environment at risk. It is also not possible to implement exemplary environmental stewardship in a hungry world where poverty is rife. It is not possible to maximize all three – despite the irrefutable desirability.

Striking the balance between the economy, environment and society requires choosing priorities and making trade-offs.

Which of these three elements matters most? My instinctive answer is 'environmental protection', but this is based on my presumption that the resources for me and my close family are secure. To concentrate on protecting the environment when the population is dying of starvation would be inhumane and impossible to enforce without coercion. The general answer to 'What matters most?' has to be 'social provision'.

Which is the biggest risk? This is where my instinct is correct. It is with the environment that we are taking the greatest risks. We can recover from financial crashes surprisingly quickly. Broken social systems can be rebuilt over a generation or two. But it is not so easy to backtrack on our environmental mistakes. Humankind now has the capability to wreck the environment so effectively that it would take thousands, or tens of thousands, of years to recover. This is not a risk we should be running.

Of the three elements, my brief analysis indicates that economics has the least direct importance. But economic measures are the most easily controlled, being specific, quantitative and measurable. They are also the prime tools we have been using to run society over recent decades, so we are familiar with them. Economic strength is the best starting point for building a sustainable society, and economic tools are some of the most effective methods we have to drive implementation.

A country, organization or individual with sound finances can afford to look beyond economics. People who live in such countries have the luxury of being able to look at life in its wider context. Those of us in the developed world with employment and secure finances can afford to be tolerant, compassionate and fair-minded. It is the tough financial discipline under which our society operates which permits this. We can learn a sustainable approach, not by diluting our economic policies or blunting the mechanisms of the market, but by bringing the same rigour to environmental protection and social provision.

Laying the Foundations

Setting the balance between economics, social provision and environmental protection requires power and commitment. Power is the capability to get something done and commitment is the willingness to do so. These two characteristics are distributed in different measures.

The government of a sovereign state is where the most power resides: in some countries, this is quite literally the power of life or death over its citizens. Paradoxically, power diminishes at the world level. The UN is, in principle, the most powerful organization in the world, but in reality it has a very loose grip on the levers of power.

Commitment is distributed according to a different profile. It is at the local level that a feeling of belonging, a sense of community and commitment to the common good are strongest. This then steadily weakens as we rise up through the administrative structures of society to world level.

At the lowest level, within a family, commitment is often absolute and the willingness to act for the common good is boundless. The influence of close personal relationships also extends into the immediate local community of neighbours and families with same-age children. Local government and local organizations are less personal, but coordination is still through people who know each other, so the complex give and take of maintaining a healthy sustainable community has a high chance of success. There will be disputes, arguments and conflicts, as there always are in human affairs, but these take place within social structures capable of resolving a sustainable balance.

This close affinity with fellow citizens extends as far as the shared need to cooperate. In the Australian outback, where the closest neighbour might be four hours' drive away, the sense of community covers a large geographical area. Within a city, the sense of community is often confined to a small area such as one apartment block, a few shops and a place to meet such as a bar or café.

A feeling of kinship can extend to the whole country if it is a nation state, as many smaller countries demonstrate. Larger countries, too, can behave as one community where commitment to shared values is strong, such as the United States.

The sense of community can also extend to groups of countries, leading to regions where many values are shared. The Arab nations are one example of a shared identity. Europe is another, although the strength of shared European identity seems to wax and wane. This has been particularly evident during the process of enlargement, as more diverse peoples and cultures are brought inside the structure of the European Union (EU).

At the global level, there is the least commitment. Very few people would regard themselves as world citizens with a moral responsibility to the world community as a whole.

This brief analysis of power and commitment leads to two important principles. The first is the principle of subsidiarity,[6] in which people understand their own environment, know the people around them, can take decisions for the collective good and take ownership of the result. The responsibility to run a sustainable society should, therefore, reside at the lowest possible level, with local solutions taking precedence over national solutions, national solutions taking precedence over regional solutions and regional solutions taking precedence over global solutions.

The second principle is the primacy of the state. This is where the combined value of power and commitment is greatest. Governments have power over borders, legislation and fundamental areas of policy and can call on a sense of nationhood, loyalty and pride. They, therefore, have the most influence over building a sustainable society.

The reality is that the world has always been run in this way, but many aspects of the policy of globalization conflict with this natural order. In defining a policy that can take us forward and bring policy closer to people's needs, I find that coining a new term is useful. This will ensure that our thinking does not slide back into the familiar furrows of existing terminology. For example, I argue for an increasingly localized economy and society, but this is not the same as the 'localization' (Hines 2000) or 'economic localization' (Woodin and Lucas 2004), described by some green campaigners, although there are a number of similarities. The new term I use is 'proximization', which I define (McManners 2008) as:

> *Proximization is selfish determination to build sustainable societies, aimed at social provision and driven by economic policy, whilst minimizing adverse impacts on the environment.*

Proximization is a natural rebalancing of the world order to return to a stable and effective world community. It will come about because of people's determination to defend their communities and ensure a safe and secure livelihood for their family and those around them. I propose that, in order to implement proximization, the following principles should be adhered to:

- decision making on the basis of sustainability – balancing the economic, social and environmental consequences;

6 This was originally a Catholic social principle that states that societies should not interfere in areas where families can decide on their own.

- subsidiarity – control left at the lowest possible level;
- the primacy of the state – where power and responsibility reside;
- use of market economics – constrained to fit local circumstances.

The adoption of these principles will lead to localizing activities and closing off process cycles within the local area as the only sure way of being sustainable. There will still be loops of activity extending out to regional and global level when this is the most effective and sustainable solution, particularly when it consists of know-how and expertise. There will also continue to be commodity flows, but on a smaller scale and operating under close oversight as we seek to influence protection of the environment in places beyond our direct control by using our power to choose what we import.

Proximization is not without difficulty, and it is likely to be a diffusion of ideas from the bottom up rather than the top down. It will take time for the changes in culture to evolve. It may be that proximization will only gain widespread acceptance when a significant cohort of countries adopts these principles, and shows that they work.

There are also risks. A half-hearted attempt at proximization could transform into a negative form of nationalism exploited by the far right. This is why we need to put a strong focus on the social element of the changes we make to ensure that we do not create an underclass where extreme nationalist ideologies can take root. Proximization requires a robust coordinated approach to delivering social benefits for the entire population of a defined community or country.

Economic Crisis Leading to Action

The deep recession of 2008 forced world leaders to re-evaluate their policies, but their focus was not on dusting off and re-reading the 2005 *Millennium Ecosystem Assessment*. Their focus, understandably, was on the short-term challenge of rescuing the world economy. Inevitably, they reached for familiar policy options to give the economy a boost: fiscal stimulus through tax cuts, spending on infrastructure and reducing interest rates to historically low levels across the world. The longer-term response required is to look beyond the economy to the whole world system and make substantive efforts to act on the warnings in the *Millennium Ecosystem Assessment*.

When the economy is booming, we tend not to question the assumptions we use in setting policy. Whatever it is we are doing, it must be right – just look at the figures. More and more countries are drawn into adopting similar policy, not wanting to be left behind. As the bandwagon gathers pace and momentum, everyone benefits. This is classic bubble dynamics, but applying not to just one industry, one sector, one economy or one group of economies. In the interconnected globalized world, world leaders in every capital city from Washington to London and from Moscow to Beijing have been enjoying sharing in collective economic success. It should come as no surprise that when the bubble bursts, we are all dragged down together. But in 2008 we were surprised at the speed with which the world's finances unravelled.

There is a danger that the focus on reform will concentrate on strengthening the current system: to give the International Monetary Fund (IMF) more power, widen the remit of the World Bank and increase pressure to finalize a new world trade agreement. This would be a mistake, and it would deflect effort that could be applied in better ways. The uncomfortable truth is that the narrow focus on short-term economic outcomes that has dominated policy choices of recent decades has been shown to be dangerous. It is dangerous to the long-term health of economies and to the integrity of the environment, and it also runs the risk of destabilizing world society as people feel they no longer have control of their own affairs.

The world economy has benefited from periods of globalization before. In the eighteenth century, there was a huge expansion in trade brought about by the technology of fast sailing ships opening up trade routes to the east. The Industrial Revolution brought on another global economic boom. After the First World War, the golden age of the 1920s ended with the Great Depression of the 1930s. I find it worrying that politicians and economists look back to previous recessions for clues as to how to respond. The current phase of globalization is different to any that has gone before. The other periods of global economic expansion were limited in scale and in the impact they had on the planet. Twenty-first century economic globalization has reached a scale that puts the planet at risk.

Finding the Moral Courage to Act

There is something fundamentally wrong with the way we are managing the world. One group of policy makers pursues a narrow focus on economic success, and is listened to, whilst in parallel another group that is seeking to safeguard the ecosystem is largely ignored. There are few world leaders with a foot in both camps, and who are willing to make substantive efforts to achieve both outcomes simultaneously. Al Gore is one of the few. Whilst serving as US Vice President to Bill Clinton (1993–2001) he published *Earth in the Balance* (2000). This showed his environmental credentials, but despite his influence and power there was little discernible movement by the United States towards policy in support of sustainability. Now outside government, Al Gore continues to press his message in the book and film, *An Inconvenient Truth* (2006). At the World Forum on Enterprise and the Environment 2009,[7] he presented two questions that we might be asked in the future. The first was, 'How come you did nothing about climate change?' The second was, 'How did you find the moral courage to rise up and solve a crisis that that so many said was impossible to solve?'

This illustrates the dilemma the world faces. Consciousness of the climate crisis is growing, but it will need to cross a threshold before politicians have the mandate to act. Climate change is a crisis, and people will accept it as such when the effects are evident and affecting everyone. There will then be a much greater appetite for reconfiguring world society according to the paradigm of sustainability.

Reconfiguring Society

As we try to define the nature of a sustainable world, and identify the levers to take us there, we have to accept that globalization is fact. Whether you support globalization, or oppose it, the world has changed. Many people campaigning against globalization seek to revert history to some point in the past. This is not helpful; we are where we are and we need to move forward from here. The changes required are fundamental. The society we now have has to be transformed. We need a bold leap forward, not defence of an old order.

7 The World Forum on Enterprise and the Environment 5–7 July 2009 was organized by the Smith School of Enterprise and the Environment and held at Oxford University, Oxford, UK.

Under the stewardship of the state, a sustainable society consisting of cooperating communities can be built. Resistance to the movement of people, goods and commodities between countries will arise, but countries also have different climates, resources and skills. Countries that adopt sustainable thinking and respect collective arrangements to protect shared resources, such as the oceans and the atmosphere, will cooperate closely. Where it is sensible to exchange goods, services or commodities, within a sustainable framework, they will do so. Countries that decide not to adopt sustainable thinking should expect to be excluded.

If the countries of the world do not take close control of their own societies – using the principles described here as 'proximization' – and rely on the open globalized world to deliver economic growth and wealth, then global governance will not be strong enough to prevent a trashed world environment that we will all have to share.

Proximization is the alternative. People living in countries that adopt it will have the luxury of observing how other countries are faring, whilst being insulated from many of their problems. Countries looking in and seeing the evident success of proximization are likely to be converted, too.

A world in which proximization takes hold will reconcile global forces with sustainable policies derived from local circumstances. We should expect a variety of economic models in the same way that we will continue to have different social models and a variety of core values for each society. The world will continue to be a rich and diverse mix of cultures and nations.

Prosperity without Growth

Conventional management of the economy is fixated on growth. Progress makes our economic activities ever more efficient. To maintain employment, we have to grow consumption to match these efficiency gains. The experience of managing economies using conventional economic tools is that growing economies are stable whilst contracting economies are not.

Recent studies have explored the possibilities of managing resilient economies without growth. The arguments can be complex and the models even more so. Peter Victor, a Canadian economist, has developed an economic model that supports low-growth economic management (Victor 2008). In the model, one aspect of a resilient low-growth economy is a reduction in working hours across the economy (Jackson 2009). At the dawn of the age of computers in the 1950s, there were predictions that we would lead lives of leisure. That was not the outcome: we chose to take the efficiency gains as increased consumption. Perhaps the time has come to compensate for efficiency gains by working less rather than growing consumption.

Low-growth management of the economy requires that a government has the flexibility to manage its affairs and to make choices that might conflict with conventional economic models. Choosing to exploit the benefits of globalization can put the government in a straightjacket. By replacing globalization with proximization, governments have some insulation from the coarse economics of open markets and free trade and have more freedom to achieve the complex balance of managing society sustainably.

Measuring Progress

Common sense says that overall world GDP would fall as policy shifted from a narrow focus on economics. GDP may well drop for a government that is successful at delivering social provision to its population and safeguarding the environment. This should not necessarily be seen as a problem, unless our actions to implement sustainability stifle innovation and breed complacency and inefficiency.

The synergies and efficiencies available from the global market will still be useful, but the context will move from pure economics to a focus on building and supporting communities. It is a fact of globalization that when borders have opened and barriers have been removed, trade has flourished. This is, of course, the intention of the policy. If one looks beyond the policy of free trade, to its consequences for social systems and the environment, it becomes clear that focusing policy in support of globalization is too simplistic. A world trade system is needed that can work in support of social outcomes in a way that safeguards the environment. Trade will adjust to match real needs in a sustainable manner between countries where different economic and social priorities have been set. Quantity of trade is a poor measure of progress. It can be argued that, in a sustainable world society, such a measure could work in reverse, with progress being defined by reductions in the international flows of physical goods and commodities.

Politicians and their economic advisors have, for a long time, focused on GDP on the assumption that this works as a proxy for progress. This assumption needs to be re-examined. Recent research has shown that as countries get richer, the assumption is no longer valid. Diener and Biswas-Diener (2008) report that increases in average individual income correlate with measures of quality of life up to an annual income equivalent to $10,000; the relationship no longer applies with increases to income beyond this level. We get richer but we do not get any happier.

The time has come to stop using GDP as the focus of policy. The world does not need a stimulus to consume and expend more. The system needs to be changed to consume much less. The requirement is to decouple human progress from environmental degradation. This is not simple, or easy, and it requires a complete change of mindset by policy makers.

There are examples of where measures of GDP are not that useful. The World Bank defines extreme poverty as living on less than US $1.25[8] per day, and moderate poverty as less than $2 a day (World Bank 2009a). An average figure for a region can be derived by dividing the GDP by the population. Although mathematically correct, it tells us little. Whilst working in Africa, I came across communities in which the paper money I carried had no perceived value. Some of these communities were exceedingly well run and seemed quite advanced in their social structures. One village I visited had no monetary income, so by economically based measures they were in abject poverty and in need of help. But it was I – with my Land Rover completely bogged down to its axles in a remote area – who needed help.

The whole village turned out to help pull me out. The only payment they would accept was a packet of biscuits I was carrying – which I supposed seemed different and exotic. I did not get the impression that this was a community in crisis. From the well-ordered look of the village, they seemed to have sound administration. From the state of the beautiful natural scenery, I could assume that they were in tune with their environment. From their laughter as they dug me out, they were, I guessed, quite content.

Transposing Western economic measures on to all the world's communities does not provide a measure of the quality of life. Measurements of social provision and environmental conservation may be more appropriate. Ultimately, we should be measuring health, happiness and fulfilment, but this is hard to

8 2005 PPP.

do on an objective basis. I leave it to others to work out what might be the best parameters, but we can be sure that GDP will be a poor measure of how we are doing when it comes to building a sustainable world. Governments should move away from using growth in GDP as their prime measure of success. GDP is, indeed, likely to continue to grow but it is the wrong measure and the wrong target.

We have choices about the future direction for globalization. We do not have such a choice with regard to a sustainable human society. This is something we have to achieve. If we fail, the ecosystem will fail and the knock-on effect will be the destruction of human society and civilization as we know it, as I describe in *Victim of Success* (McManners 2009). The only uncertainty is the timescale.

I believe that it might be another two decades before climate change causes dramatic impacts on human society. Oil may last for a similar time. We cannot afford to wait that long in complacent ignorance. The changes that must be made to society and infrastructure are so fundamental that they could take 30 years to implement. Add to that a delay of 20 years, and it will be 50 years before we have brought society back into balance with nature. Only an optimist (or a deluded fool) could believe that the Earth can withstand yet another half century of sustained industrialization.

The time has come to deliver green outcomes in the real world. We must use determination to take control of local conditions to restrain the global forces that threaten to destroy us. First, we need to understand the concept of sustainability.

2

Sustainability

A sustainable policy framework requires a balance between social provision, safeguarding the environment and sound economics, described neatly as 'people, planet, profit'.[1] The World Commission on Environment and Development (WCED) report, *Our Common Future* (WCED 1987), defined sustainable development as, '… development that meets the needs of the present without compromising the ability of future generations to meet their own needs'.

This has since been adopted widely as a definition of sustainability. It was an appropriate definition at the time of writing and it suited the purpose intended. The report, which has become known as the Brundtland Report after the commission's chairman,[2] was concerned with securing global equity and redistributing resources towards poorer nations whilst encouraging their economic growth.

During the decades since the Brundtland Report, the scope of sustainability has expanded and the imperative to find sustainable ways to run society has become ever more urgent. Barry (2007) argues that to implement the principles of sustainable development requires a 'clear shift towards making the promotion of economic security (and quality of life) central to economic policy'. Many green economists go further and now regard sustainable development as 'an oxymoron, in reality often counteracting existing, local and community economic patterns' (Kennet and Heinemann 2006). The word 'development' is no longer suitable for inclusion in the core definition of sustainability. It brings with it the assumption that development is inevitable. Lunn (2006) believes that the notion of a sustainable society must change: 'Developed countries must

1 The phrase was coined for Shell by the consultancy SustainAbility in 1994 and was widely credited to John Elkington (founder of SustainAbility).

2 Gro Harlem Brundtland chaired the WCED, which was convened by the United Nations in 1983.

recognise that there are alternative models to rampant economic growth while developing countries must invest in capacity building in order to increase their self-reliance.'

In the second decade of the twenty-first century, a more accurate definition of the objective of sustainability is to secure the future of civilization. Until recently, we assumed that whatever policies we adopt, civilization will continue. This is no longer a safe assumption. A blinkered focus on development is one of the misconceptions we must now put aside. I propose a more appropriate definition of sustainability:

> *Sustainability is the concept of securing the future of human civilization concurrently with safeguarding the integrity of the Earth's ecosystem.*

My definition may be overly dramatic for some people, but in my book *Victim of Success* I laid out a credible scenario that the future of civilization is at risk. The starting point for real sustainable policy is to understand and accept that this unwelcome analysis may be true. From this mindset we can short-circuit the politically correct approach of attempting to solve every one of the world's problems to focusing on the prime parameters of a safe future.

The Club of Rome pointed out the dangers in its report *The Limits to Growth* (Meadows 1972), which modelled the consequences of growing world consumption linked with finite resources. The report was controversial and much criticized. Its message was not liked by policy makers and was largely ignored, as the authors have acknowledged (Meadows et al. 2004). The report predicted that, without major changes of policy, the ecosystem and world economy would collapse in the middle of the twenty-first century.

Recent analysis in 2008 by Graham Turner of Australia's national science agency compared the predictions made in *Limits to Growth* with 30 years of actual data. He found a close match between the report's 'standard-run' (business-as-usual) scenario and the historic data for 1970–2000 (Turner 2008). This would seem to indicate that the predictions in *Limits to Growth* were fairly accurate. The central assertion that there is a limit to growth that cannot be exceeded is hard to dispute.

During the four decades since the *Limits to Growth* report, all of us are guilty of not responding. Economists are at fault for failing to champion a new approach, even though no- or low-growth economic management is feasible

(Daly 1991a). People everywhere are at fault for their reluctance to cure their fixation on material advancement, even though they know it does not make them happy. Supporters of sustainability are at fault for giving too much weight to global equity, concentrating on a perceived lack of development instead of the integrity of the ecosystem. In this chapter, I concentrate on bringing sustainability back to basics. As I repeat throughout this book, economics is a tool of policy; we have to get the policy right before applying economic methods. Understanding sustainability is one prerequisite for setting effective economic policy.

The Rise of Sustainability

The foundations of the current movement calling for a more sustainable world date back to the declaration that was agreed at the Earth Summit in Rio de Janeiro in 1992. Central to the discussion of poverty, war and the growing gap between industrialized and developing countries was the question of how to relieve the pressure on the global environmental system through the introduction of the paradigm of sustainable development.

The Rio Declaration (UN 1992) states that:

> *Human beings are at the centre of concerns for sustainable development. They are entitled to a healthy and productive life in harmony with nature ... In order to achieve sustainable development, environmental protection shall constitute an integral part of the development process and cannot be considered in isolation from it ... Peace, development and environmental protection are interdependent and indivisible.*[3]

The Rio Declaration is an important statement of intent. It is backed up by the Commission on Sustainable Development and an action plan called Agenda 21. The principles were reaffirmed in 2002 at the World Summit on Sustainable Development (WSSD).[4]

Despite the good intentions and considerable effort by the Commission on Sustainable Development, and others, very little substantive progress has been made. The UN is constrained by the need to incorporate a wide range of opinions: the full 27 principles of the Rio Declaration cover every conceivable

3 This extract consists of Principles 1, 4 and 25 of the Rio Declaration.

4 The WSSD was held in Johannesburg, South Africa, 26 August–4 September 2002.

concern. Real action will come from governments and other organizations that have more freedom of choice and freedom to act. The UN has articulated some high-sounding objectives. The only way to make progress towards these is to translate them into a message that the people with power in politics and business can embrace.

To begin the process of translating the concepts of sustainability into sound policy, it is necessary to examine the views expressed by our global institutions. An examination of these aspirations has to be brought into the argument because unrealistic expectations at this level are a barrier that has to be overcome.

Unrealistic Expectations

The *Millennium Development Declaration* (UN 2000a) and the associated *Millennium Development Goals* (*MDG*) (UN 2001) encapsulate the development aspirations of the world community. The *MDG* encompasses human values and rights such as freedom from hunger, the right to basic education, the right to health and a responsibility to future generations. Goal 1 is to 'Eradicate extreme poverty and hunger' with the prime target of halving the proportion of people whose income is less than one dollar a day by 2015 compared with 1990. There then follows goals for education, gender equality, child mortality, maternal health and combating disease. Down in seventh place, out of a list of eight goals, is 'ensure environmental sustainability'.

The *MDG* is a 'human-centric' plan. Each goal or target, taken in isolation, is clearly beneficial to humanity. Taken as a whole, inherent contradictions have been allowed into the process. Compromise is needed between different targets and goals to come up with a feasible plan. We need to pick out the key targets that must be achieved for a safe future for humanity and, whatever else we do, hit those key targets without fail. This is how action takes place in the real world. The secondary targets are then met in so far as is possible, as long as the primary necessary action is not undermined.

Concentrating efforts on a basket of medium-term targets to improve human welfare runs the risk of inflicting far worse human suffering in the long term. It is difficult to argue that some of the gaols are less important than others because someone will suffer as a result. However, it is acceptable to single out the key target (MDG Monitor 2009):

> *Target 7A: Integrate the principles of sustainable development into country policies and programmes and reverse the loss of environmental resources.*

This is the target that matters more than any other, because it is the one that underpins the possibility that we might achieve the others. We can put effort into this target knowing that the consequential effect on other targets will be positive.

> *It is now our responsibility ... to put all countries, together, firmly on track towards a more prosperous, sustainable and equitable world.*
>
> *Ban Ki-Moon, (UN 2008)*

This is the sort of statement we expect from the Secretary General of the United Nations, and of course he has wide support. This is what we want to hear, but it is not what needs to be said. If the world really wants to break out of the business-as-usual scenario, we need leaders who are willing to tell some hard truths. The world faces a crisis. The worst effects might still be four or five decades away, but the only way to stop humanity from colliding with the limit to growth is to start taking action now.

Once it is accepted that there is a pending crisis, the focus must be on the single key issue of sustainability. Prosperity and equity will have to wait. A sustainable world may be less prosperous in the short term, but an unsustainable world will, over time, destroy prosperity so completely that recovery in any sensible time frame will be impossible. The sustainable world that I envisage will not be free of problems. The inequity observed at global level will continue, but this must be balanced against the reinforcement of equity at the local level – this being a natural consequence of sustainable policies. Sustainable solutions do not lead to Utopia, but the overall result is better than any other policy framework. World leaders should not be frightened to push past the political difficulties to focus on sustainability and see what sort of policies it leads to.

Adopting the Concept of Sustainability

The theory of sustainability is based on the three mutually reinforcing pillars of the economy, society and the environment. It is becoming widely accepted that the economy cannot be considered separately from social and environmental issues, remembering that the converse is also true. Building a fair society and protecting the environment must be based on sound economics.

The route to a sustainable world will not be easy. The developed world will be reluctant to turn its back on the economic progress that has been made over the last century (and the associated lifestyle). The underdeveloped world will not want to forgo its chance of an industrial revolution to pull its populations out of poverty. It will not be easy to persuade either the rich or the poor to change in our struggle to find pragmatic policies with the potential for sustainability in the real world. Action is required by all peoples and all countries, but the world community has, so far, failed to find a way.

There is something missing from the model of sustainability we have been using over the last two decades. We have failed to appreciate that the three pillars of a sustainable human society have foundations on the earth beneath our feet. The area of all the continents of the world is a fixed constraint. This is something that we cannot change and we have to live within the land's capacity.

Land – the Foundation of the Ecosystem and Society

There are three main categories of land use: urban, agricultural and nature, listed in the order of the economic value we give them. In our economic system, the land that we decide to leave for nature is valueless (unless it is sufficiently unusual and interesting to attract paying tourists). Converting natural wilderness or forest into agricultural land generates economic value. Cash crops can be grown and loans for machinery secured against it. The uplift in value is considerably higher for land converted to urban use. Landowners looking for the best economic return should seek to build houses or commercial buildings. Purchasers will pay a good price and tenants a good rent. A capitalist system underpinned by conventional economics encourages landowners always to be on the lookout for uplift in value.

It is in the economic interest of owners of wilderness to look out for a chance to clear it. The owner may have other personal values and intentions, leading away from a focus on their economic best interest, but in our current economic system there is always an entrepreneur waiting to take control. If a way can be seen to navigate through the planning rules, there is cash to be earned. In a country with a robust cadastre,[5] tight planning controls and sound governance, the opportunities are limited. Even so, whenever a change of policy allows land in its natural state to be taken into the economic system, someone is waiting to pounce on the opportunity.

In countries with weak governance, the economic pressures are harder to resist. Often these are the same countries that own the largest remaining wilderness areas. When land-use policy is the de facto acceptance of whatever happens on the ground, or it can be amended by payment of the appropriate bribe, there is no effective protection for the land required by nature.

Capitalism and conventional economics lead us to chip away forever at the land given over to nature. It is only national parks that are protected by robust national law in countries with good governance where nature will find a long-term home. There is no financial incentive to return land to its natural state, so it is rare that land is reclaimed from urban or agricultural use. The exception is land that has been so badly polluted that it cannot grow crops for human consumption and no one is willing to erect buildings on it. It has then become worthless, so may be allowed to revert to nature.

The fact that we have an economic system that does not value the land we allocate to nature is not just a curiosity – it is highly dangerous. The ecosystem of the planet needs land to operate. Ideally, the land taken over by humankind would be small patches cut out of the systems of nature, thus allowing the ecosystem to carry on much as before. This was the case throughout human history until relatively recently. There have been vast areas of the planet left for nature: the Amazon rainforest and much of the continents of Africa and Antarctica (although the contribution of the cold continent to the global ecosystem may be limited). In Europe, much of the original forest has been cleared, though efforts are made to manage and protect the 35 per cent that remains (Vanhanen et al. 2007). The traditional European landscape includes small patches of nature retained at the micro level, for example hedgerows. As the pressures of capitalism and the free market are played out, the last great rainforests are being cleared to grow soya beans or crops for biofuel. In Europe, hedgerows are being ripped out to improve the economic efficiency of farms.

We need to adopt a policy model in which land is the foundation of the ecosystem and society.

5 A cadastre is a comprehensive land register including details of ownership, tenure, precise location and area.

A Model of Sustainability

Sustainability requires striking a balance between human needs and the ecosystem in a way that can endure into the long future. The model I propose has at its base the three categories of land use: urban, agricultural and the land left for nature. These must be in equilibrium. The land we take for our cities must leave enough for agriculture to feed the population, which in turn must leave enough for nature to retain the integrity of the ecosystem.

The sustainability of land use underpins the sustainability of human society. This foundation policy is 'Earth-centric' with the focus on ensuring the integrity of the ecosystem. If we believe that the focus of land use should be elsewhere, such as increasing the amount of land we allocate to agriculture in order to be able to feed a growing human population, then the precondition is that enough land is left for nature to survive intact into the long future.

Above this foundation, a sustainable society requires a balance between three areas of policy: the environment, the economy and social provision. This balance should be applied to all decisions, particularly major decisions with impacts that will endure into the long future. Such decisions cover a diverse range from infrastructure planning and development to education and social welfare policy. This is 'human-centric' decision making, in which social provision is paramount.

There is an inherent contradiction between 'human-centric' sustainability, focused on social provision and 'Earth-centric' sustainability focused on the integrity of the ecosystem. We want to achieve both, of course. Where this is not possible, securing stability of land use must take precedence. When the stability of the ecosystem is at risk, land-use policy must take priority in searching for solutions. This will then limit the options available in setting sustainable policy for society. There may be some very tough choices, which entail social hardship and/or tough economic measures, but this is the price of long-term sustainability for society and the planet.

This model of sustainability can be depicted as two stools standing one on the other, as shown in Figure 2.1.[6] Standing on just one stool is very stable. When the stool we are standing on is perched on top of another stool, the situation is much more precarious. Instability in either stool could tip us off

6 I introduce the physical analogy of two stools standing one on top of the other in *Victim of Success* (McManners 2009).

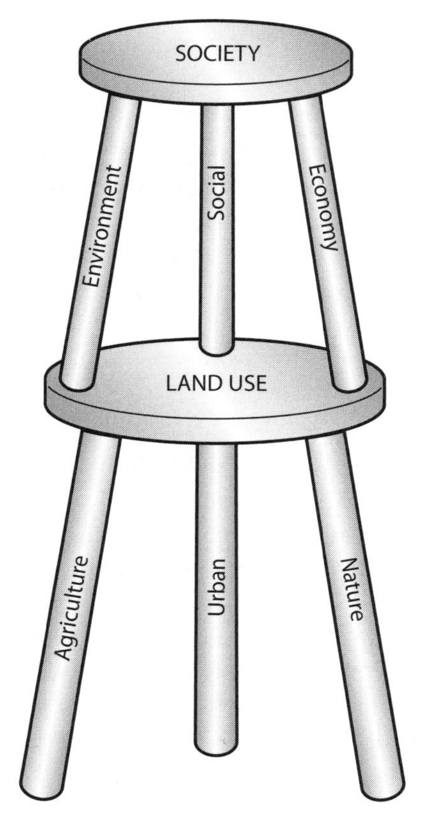

Figure 2.1 The Two-Stool Model of Sustainability

balance. Instability in both stools makes balance almost impossible.

When both stools are in balance, it is a robust and stable system. If the three legs of society get out of balance, with one leg much longer than the others, or one much shorter, society becomes unstable. If the imbalance continues to worsen, society will collapse. Zimbabwe in the first decade of the twenty-first century is an example. A country with wonderful natural resources, and which used to be the bread basket of Africa, Zimbabwe has been afflicted by gross mismanagement. Zimbabwean society has collapsed and the population is suffering from hunger and easily prevented diseases such as cholera. In this case, it is society that has become unstable. An imbalance in land use is not the problem. Mistakes can be fixed. Zimbabwe has a benign climate, fertile soil and wonderful people. With better political leadership, supplemented by help from international donors, the economy can be rebuilt. It is reasonable to be confident that Zimbabwe will recover. The upper stool in the model, representing the local society, has toppled over. With a little help from outside the country, Zimbabwe can be returned to a safe and stable condition.

When we have problems with both the sustainability of society and a growing imbalance in land use, we could have a dramatic collapse. For example, as we approach the end of the era of cheap oil, two legs of society are coming under pressure: we face economic hardship together with the environmental problems of climate change. At the same time, land use is coming under pressure to change. To help solve our problems, we are pushing to allocate more land to grow crops as feedstock for biofuel production to reduce our reliance on oil. This conflicts with the requirement for food. We want to allocate more land to agriculture to be able to continue to feed a growing population. The knock-on effect is a rapid increase in the rate of deforestation and the destruction of natural habitats. The 'nature' leg of the bottom stool is being dramatically weakened as more and more land is claimed from nature.

There is a real and present danger that the foundations upon which we are attempting to build sustainable societies are crumbling. The foundations of land use at the local, national, regional and global level must be stabilized as a matter of priority. Understanding this model of sustainability is fundamental to finding effective methods to apply sustainability policy in the real world.

Principles of Sustainability

The concept of sustainability has been used to cover a range of objectives, so there is no widely accepted set of principles, as Simon Dresner (2008) found in his review of the literature. Dresner quotes four principles of sustainability proposed by Herman Daly (Daly 1991b). These can be summarized as: limiting human activities to be within the Earth's capacity; ensuring that technological progress increases efficiency rather than throughput; harvesting renewable resources should not exceed regeneration rates; and exploiting non-renewable resources should proceed no faster than the creation of renewable substitutes. These are useful policy guidelines but they are not sufficiently resolute to serve as principles.

The principles I propose to underpin policy are consistent with my definition of sustainability (see p. 22):

1. Sufficient land must be left for nature to maintain a stable ecosystem.

2. Social provision and safeguarding the environment must take precedence over economic outcomes.

3. All development must include elements that reinforce the ecosystem.

1. Sufficient land must be left for nature to maintain a stable ecosystem

The first principle of sustainability is to retain enough land in its natural state to ensure the stability of the ecosystem. We cannot know how much land this is. Ideally, we would have confiscated such a small proportion for agriculture and urban dwelling that there could be no doubt of the dependability of the ecosystem. This is how it has been for most of human existence. It is only recently that the dramatic expansion of the human population, together with expansion in consumption, has been threatening to destroy the ultimate foundation of society. In the developed world, there is limited land left for nature. We must adopt uncompromising and robust policy to defend what is left and work at encouraging nature back into sharing our agricultural land and even into our cities. We must also work very hard to influence other countries to protect and defend the land allocated to nature outside our borders. This leads into the second principle of sustainability.

2. Social provision and safeguarding the environment must take precedence over economic outcomes

In a sustainable world, the prime policy objectives will be social outcomes, backed up by robust measures to safeguard the environment. The role of economics is to support these objectives. Economic tools are useful because they are quantifiable, measurable and we are familiar with the methods. The danger comes when economic outcomes are set as objectives. This is justified on the assumption that economic progress and human progress are closely aligned. It is now known that this is not necessarily true. Economics is an important tool to facilitate achieving policy objectives, but it must be subservient to the other two.

This principle will become superfluous once sustainability is established as the core principle of the management of society. Until then, specific effort is needed to break out of the trap that economic outcomes trump other policy.

3. All development must include elements that reinforce the ecosystem

Human impact on the planet is significant and leading to dramatic changes. The forces of nature and the cycles of nature used to be in balance. Anthropogenic forces are now upsetting the balance and we do not know for certain what the outcome will be. These are dangerous times. Minimizing the environmental impact of a proposed development, process or project was appropriate policy when the overall impact of humans on the Earth's ecosystem was small. This is no longer enough. Each time a major project is planned, the end result must be a net improvement to the ecosystem. Over the last half century, each major project has destroyed a little bit more of the Earth's ecosystem. Each new human facility such as a new town, industrial estate or harbour eliminates another small piece of nature. No one project can be blamed, but the sum total is a steady decline in natural capital. The process can be reversed using the same project-by-project approach, but this time to add back natural capital.

Global Sustainability

Sustainability is the concept of securing the future of human civilization concurrently with safeguarding the integrity of the Earth's ecosystem. In history, civilizations come and go. The Roman Empire, the Mongols, the Maya, the British Empire were all successful in their day, but history tells us that collapse is inevitable (Tainter 1990). There is something about the human condition that prevents us from establishing societies that are stable indefinitely. It might be innovation and curiosity that make us look for change, or corrosive corruption brought on by success that destroys a civilization from within.

The process of globalization, if run to its natural conclusion, would lead to a single global civilization. This should be deeply worrying. Humans have never before constructed a stable society. If we are constructing a global civilization, the lessons from history are clear: it will collapse. Therefore, a safe future for world civilization must be as a patchwork of different societies. As one starts to struggle, there will be others to take up the lead.

At the global level, the biggest challenge of sustainability is safeguarding the integrity of the planet's ecosystem. The whole of humanity shares the atmosphere and the oceans. They interconnect as one biosphere. Pollution and damage caused anywhere across the world affects us all. Land is also a vital component of the shared biosphere, providing ecological service such as forests that absorb carbon dioxide. Civilization relies on ecological services and it is vital that they are protected.

The world needs enforceable global agreements for the protection of the shared biosphere. Negotiating these requires that global institutions and global forums of world leaders use the logic of sustainability in their decision making. This is not happening. In particular, world leaders and national representatives negotiate with a focus on their best national interest. Giving ground in negotiations to achieve a sustainable global solution does not come naturally.

The self-interest that dominates international relations is not going to change easily. If we put our trust in world institutions to protect the ecosystem and wait until we can achieve effective world-level agreements backed up by watertight enforcement, we will be waiting for a very long time.

The evident self-interest in a stable ecosystem, and the urgency of protecting it, leads towards unilateral action. We must take action within our own jurisdiction and area of control to enforce sustainability. The leaders of the developed nations have to recognize that poorer countries want to follow our example. But there is a lead time before they reach the stage that we are at now. It is vital that we move very fast to parade a different example. In addition to providing self-interested leadership towards a sustainable society, developed nations should wield the stick of enforcement in so far as their power and influence extends. Developing nations, too, should act within their own territory to become sustainable. Whilst the West is learning to become sustainable, developing nations should take note that the Western model of consumption is flawed and be very wary of advice coming from the World Bank until the paradigm of sustainability is deeply rooted in World-Bank policy.

Sustainability is a simple headline concept, but it is completely different to the thinking that underpins the policies of economic globalization. It will take time and effort to establish new thought processes, and for these to become widely ingrained. I am optimistic that, over the next two decades, this can be done – but the current generation of leaders will resist. They have experienced the economic benefits of twentieth-century policies and will be reluctant to shift direction. The new generation coming into power will have the experience of environmental damage to guide their thinking. Instead of predictions, they will have hard data and a number of immediate challenges, such as a surge in the number of climate refugees. Sustainability will then be an imperative, not a concept on the edge of policy formulation.

The problems of human world society are seen most clearly from the top down. The sum total of a myriad of human actions is damaging the biosphere. A myriad of improvements are required, the sum total of which will put global society back onto a safe track. The route to a solution is from the bottom up, by people adopting sustainability as the core principle to manage communities and society.

3

Subsidiarity

A greener economy must achieve a balance between social provision and safeguarding the environment. This can be done most effectively by human-scale communities where the give and take required can be brokered face to face. The winner of the Sveriges Riksbank Prize in Economic Sciences 2009, Elinor Ostrom (2009), champions this approach in her work on economic governance, particularly with regard to the commons. This logic leads to the principle of subsidiarity, in which the control of human affairs ought to be handled by the smallest, lowest or least centralized competent authority. This principle is contained in many constitutions around the world, including the United States Constitution[1] and in the founding treaty of the EU. Article 5 of the preamble to the Treaty on European Union (EU 2008a) states that 'decisions are taken as closely as possible to the citizen in accordance with the principle of subsidiarity'.

The principle of subsidiarity is fundamental to achieving a sustainable future. Subsidiarity ensures that action taken within society is effective and has the support of citizens. The special requirements of the shift to a greener economy are particularly suited to such decentralized local control. Sustainable policy requires balancing conflicting issues. At the local level, people understand the issues and have a direct personal interest in success. When people are fully involved in decision making, there is collective ownership of the solution and implementation becomes easier. The concept has an impressive record of success, ranging from military operations to the delivery of aid in Africa.

The conduct of military operations is interesting. Even in an environment of total authority, where commanders have considerable power to give orders, and expect them to be obeyed, subordinates work more effectively when the

---

1 United States Constitution, Amendment 10 – Powers of the States and People. Ratified 12/15/1791. 'The powers not delegated to the United States by the Constitution, nor prohibited by it to the States, are reserved to the States respectively, or to the people.'

principle of subsidiarity is enforced. In the military, this is called Mission Command – a style of command promoting decentralized command, freedom and speed of action, and initiative. Having been briefed on the commander's intentions, their own mission and the context of the mission, subordinates are told what effect they are to achieve and the reason why it needs to be achieved. They then decide within their delegated freedom of action how best to achieve their mission.

In civilian management, the term 'empowerment' is used. Employees need to understand the strategy of the organization and the context of their role, together with a specific outcome that the organization is striving to achieve. Employees are then trusted to get on with making their own decisions.

In poverty alleviation, adopting subsidiarity has been shown to achieve successful outcomes. The empowerment approach focuses on mobilizing the self-help efforts of the poor, rather than providing them with social welfare. Economic empowerment has improved the lives of poor communities in many developing countries across Africa and Asia. Initiatives like Grameen Bank[2] show that people can be very effective in improving their circumstances, leveraging huge social benefit from small injections of capital. Aid agencies are learning that the old model of sending in resources according to priorities and plans set by outside 'experts' may produce short-term gains but it is not the way to deliver sustainable improvement.

An economy works for the benefit of society if people and communities are trusted to make their own choices and decisions within a national policy framework. Recent research has shown that the extent to which a society allows free choice has a major impact on happiness (Inglehart et al. 2008). This further reinforces the concept that higher authority should set rules only as necessary to deal with issues that cannot be handled at the local level.

Subsidiarity and Sustainability

The world is consuming 30 per cent more than its ecological capacity can deliver (Hails 2008). This is running down the ecological reserves of the planet. This level of overspending on the ecological current account cannot continue

2 Grameen Bank was established as an independent bank in Bangladesh in October 1983. It is owned by the rural poor whom it serves. The bank's borrowers own 90 per cent of its shares, while the remaining 10 per cent is owned by the government.

indefinitely. If all the countries of the world adopted a consumption profile that matches an average European citizen, we would need the ecological capacity of 2.6 planets. If the world followed the consumption model of the United States, even four planets would not be enough. World society has to find a way to curb consumption and to set balanced sustainable policy. The principle of subsidiarity helps to achieve a sustainable balance through facilitating compromise between conflicting priorities, in a way that attracts the support of the population.

Balance

A sustainable balance needs to be found at all levels of society to protect shared resources and the ecosystem. Subsidiarity requires that this balance should be set, primarily, at the micro-level as the most effective way of achieving the macro-balance the world needs. Some people argue that sustainability at the local level is not necessary, provided we have global agreements that balance resources. This fallacy applies particularly to carbon dioxide emissions. In order to take effective measures to cure the world's addiction to fossil fuel, action is needed particularly at the local level. Any global agreement should support and reinforce such action, not replace it.

Where there are problems on a global scale, discussion is needed at the level of world leaders, leading to agreement to act. It is obvious that the implementation of any such agreement has to cascade down to the local level. According to the principle of subsidiarity, global agreements should only apply in so far as the problem affects the world as a whole. This would indicate that world policy making should roll back some of the intrusive requirements that have arisen. For example, it could be argued that some aspects of the UN Charter for Human Rights, no matter how well intentioned, go beyond what is necessary.[3] There is no pressing reason at global level to interfere in the affairs of a sovereign country simply because we disagree with their values or dislike their internal policies. The principle of subsidiarity is too important to allow it to be subverted by such issues.

In other respects, world agreements need to be stronger where there is a pressing need to balance social, economic and environmental considerations.

3 I have not attempted to identify any particular aspect of the Charter of Human Rights that is over-intrusive because that would be bound to offend someone. This is the point. The world has different cultures and, as I argue later, this is a strength that should be retained and reinforced. There is a risk that an over-detailed Charter of Human Rights becomes a blueprint for a homogenized global society.

For example, closer linkages between economic and environmental policy are needed. This is sorely lacking, producing contradictions such as World Trade Organization (WTO) rules clashing with attempts to tighten environmental regulations.[4] Global institutions have to learn to work in concert with one another. For example, the work of the International Monetary Fund (IMF) and World Bank has to be coordinated with the UNEP.

The principle of subsidiarity linked with sustainable policies leads to less intrusive but better coordinated global policy.

Material consumption

Since the twentieth century, industry has grown on the assumption that resources are unlimited. The invisible hand of the market is then used to regulate supply and demand. When there are shortages, the price rises and investment flows into developing more capacity. For the extractive industries, such as mining and oil, high prices support the business case to exploit smaller deposits in remote locations. For agriculture, if food is in short supply, the price goes up and farmers can afford to invest more to increase yields, such as making greater use of fertilizer. Marginal land can be brought into production as high prices support the business case for investment in boreholes and water pumps.

These economics have been based on a false assumption. Natural resources are finite. To be sustainable, society has to consume at a rate that does not exceed the ecological capacity of the planet. This means a reduction of at least 30 per cent on today's levels of consumption. We must not only stop the rise in consumption: we must also put the material demands of global society into reverse. The fact that this will be hugely challenging does not alter the logic that this is the action we must take. A way has to be found.

The key policy for reining in consumption is to connect people with their resource base. When a society has control of its resources, the limitations become obvious. People will discuss how to change demand to match availability. Pricing and market mechanism have a role, of course, but localizing (in effect capping) the market provides market participants with the information they need to make well-informed choices to act in advance of the coarse market mechanism of rapidly escalating prices.

4 In 1991 Mexico lodged a complaint under the GATT dispute mechanism procedure that the United States was imposing its rules on dolphin protection on tuna imported into the country from Mexico (and other countries). The complaint was upheld.

The principle of subsidiarity pushes decisions on dealing with resource limitations down to the lowest possible level. This provides not only a mechanism to regulate overall demand, but it also supports another key aspect of material consumption – recycling. This term has been hijacked to describe the limited sorting of rubbish being generated by the throwaway society. This is not the meaning I have in mind. True recycling makes rubbish obsolete and is a principle of manufacturing and consumption in a sustainable world. Every object has a full lifecycle. This protects the environment from waste and enables society to deliver human progress in a way that does not run down the ecological reserves. To deliver true recycling, we need to rely on robust local arrangements. Subsidiarity comes in again as a key foundation to recycling policy. This issue is returned to in Chapter 7 with an examination of how global commodity flows will alter in a sustainable world.

Oversight of Capitalist Forces

Laissez-faire capitalism, which is based on individuals or organizations striving for their own best interest, tends to undermine community values. We have accepted this weakness because of the positive economic outcomes that such capitalism can deliver. At family and small community level, direct human interaction transcends capitalist forces. When the principle of subsidiarity is enforced, capitalism can be held in check.

The most disaggregated unit of human society is the family. At this level, thinking sustainably is automatic. Humans are, by nature, social beings and family groups manage the complexity of their affairs through interaction with other members of the family. Some families may be exceedingly well ordered and others less so, but there is a deep-seated, almost subconscious, knowledge that there must be a balance between social, economic and environmental considerations. People want their family to be happy, well fed, prosperous and living in a pleasant environment. The individual is willing to accept considerable personal disadvantage if it benefits the family as a whole. This is particularly so with regard to parents wishing to provide well for their children.

The ability to negotiate solutions that benefit the whole family or tribe is the basis of human success. Excessive individualism would not have led to the highly complex and interconnected civilization that has evolved. The principle of subsidiarity ensures that society retains this human attribute of managing affairs for the common good.

When the father of modern economic theory, Adam Smith, championed market mechanisms two centuries ago, he was not writing about the globalized world of today. He was thinking about economically efficient ways to run society as he saw it. In his words (Smith 1776), 'It is the maxim of every prudent master of a family, never to attempt to make at home what it will cost him more to make than to buy.'

The division of labour is effective in getting work done efficiently within a community. The baker should bake bread; the doctor should treat patients; the farmer should grow food. Each specialist can carry out the activity better than others in the community. This system used to be self-regulating. Each area was subject to natural oversight by the community.

Adam Smith extended his observations to international affairs: 'If a foreign country can supply us with a commodity cheaper than we ourselves can make it, better buy it off them with some part of the produce of our own industry, employed in a way in which we have some advantage.' This is the basis of the comparative advantage of nations which has become a fundamental principle of international economics. In tandem with this, environmental and social impacts have been allowed to become externalities, particularly at the international level where oversight is weak.

In a global free market based on laissez-faire capitalism, there is no natural community oversight. The financial incentive is for every producer to save money in the production process by whatever means possible. The consumer then purchases a product based purely on price, appearance and features without concern for the means of production. Regulations are put in place, of course, but without natural community oversight these can often be ignored or bent. Adam Smith would be horrified[5] at some of the practices that have come about in the single-minded pursuit of the economic theory he expounded: exploitation of workers, environmental damage and stomach-churning practices in meat production for export to a nameless, faceless consumer at the other side of the world.

---

5 Adam Smith was a compassionate man as shown in The *Theory of the Moral Sentiments* (Smith 1759). His theories of economic efficiency were formed whilst living in a very different society. This earlier book indicates that he would indeed be horrified to have his ideas connected with a form of unfeeling capitalism.

Subsidiarity and Cities

Small communities are more amenable to the concept of sustainability. It is possible that a small town surrounded by farmland could be self-sufficient for most resource inputs. This was the normal mode of existence for most of history. The town would act as an administrative centre for the region, and the region would have sufficient agricultural capacity to provide staple supplies.

Traditional towns can work in symbiosis with their locality up to a size of about 500,000 people.[6] Cities up to this size have the ideal population for sustainable communities. Above this size, cities need to draw inputs from further afield and lose the close relationship between consumption and agricultural capacity. I do not argue necessarily that cities should be localized into completely self-sufficient entities. However, it is a concern that it becomes very hard to set sustainable policy when the population loses the connections between people and nature, and land and food.

Modern cities have expanded beyond this simple model. Some cities have expanded into huge megacities in areas such as the northeast United States, the Rhine-Ruhr metropolitan region in Europe and the Pearl River Delta in China. Cities such as Tokyo are truly global cities drawing in resources and people from around the world.

Megacities that expand and swallow up other cities to form huge metropolitan regions can be likened to a cancerous growth which destroys the natural environment and reduces the quality of life for many of the city's residents. Unchecked, it is possible to imagine such cities expanding and merging over time to gobble up entire countries.

Modern cities suffer a number of problems. Huge impersonal cities isolate the individual, leading to loneliness and allowing crime to flourish. As people are less willing to live and raise families in the city, commuting becomes the norm, leading to suburbanization. The car has been one of the biggest influences on city design during the second half of the twentieth century. Dominance of cities by cars is now one of the biggest problems faced by city policy makers.

6 There is no definitive figure for the size of an ideal sustainable city. I use the figure of 500,000 as it is large enough to support a full range of services and small enough to walk or cycle for most journeys, provided the density is high enough.

Cities for the twenty-first century need to be designed around people and their needs. The ownership of cities should be handed back to the residents by passing responsibility down to specific areas or community groups. Our cities, even very big ones, can be a tessellation of urban villages that are on a human scale, where people interact face to face with people they know to manage their affairs. Crime is discouraged by the oversight that such a community provides. Each community develops its own character. Some are better managed than others. Some attract certain types of people and commercial activities that congregate to share the benefits of clustering. Despite such specialism, the support network of people who run the community live and work locally.

Designing cities for people and passing decision making to as low a level as possible can produce radically different cities. Taken together with other improvements in clean technology and better use of public and private space, cities can provide excellent quality of life. Such vibrant and densely populated communities are ideal for supporting a large population sustainably.

Increasing the Resilience of Global Society

Global society faces some big challenges as three problems come to a head over a similar timescale: population growth, diminishing oil reserves and climate change. These are issues that all humanity shares. These issues would not exist, or would not be as severe, if we had learnt to run society in a sustainable manner and developed sustainable economic policies at a much earlier stage. This is not where the world is at. Global society owns, collectively, these big issues and they will not now respond to purely local action.

The principal global challenges are systematic problems brought on by widespread mismanagement causing severe long-term impacts. If mismanagement is confined to one community or one country, this has limited global impact. Humans will always make mistakes – it is in our nature to innovate and experiment. This ability is the basis of our success, but it can also be our downfall when our experiments are replicated rapidly around the planet.

There are very close linkages between the natural world and human society, and also close parallels in the way they operate. This is where I turn next for inspiration. A stable ecosystem requires a wide range of biodiversity. There is a dynamic balance between the species. At any one time, there will be some dominant species that thrive. Over time, conditions change and other species tend to gain the advantage. Some species become extinct, and other species arise to fill new niches in the ecosystem. There is always alternative genetic code in small enclaves waiting for the chance to expand in number. The parallel with human society is that we, too, need a wide range of types of society and alternative economic systems for a stable world society. One economic system may have characteristics that are capable of developing technology to take humankind off this planet to populate other planets and other solar systems. Another system may prove to be very good at safeguarding the ecology of planet Earth.

Ecosystems are put at severe risk of collapse by overinvesting in a small gene pool. Similarly, world society is at risk if we invest in a limited range of economic models. Global homogeneity leads to limited alternative options, meaning that any flaw in the system is hugely magnified. The United States has set an example of dynamic free-market capitalism that is hugely successful but, like all human systems, it has flaws. It is based on high levels of material consumption. It is only a small exaggeration to claim that if the whole world adopted the US economic model, the human species would, in time, become extinct. One flaw magnified at global scale becomes a huge systematic problem.

It can be argued that the solution in this particular example is to fix the flaw in the US economic model and continue to roll it out around the world. Ecosystem theory points towards a more general solution. We must not allow a single economic system to dominate. Any single human system will have flaws. For macro stability of human society we need variety.

Variety brings resilience and stability at the macro level. In an ecosystem, individual species come and go, but that is less important than the integrity of the ecosystem being preserved by the turnover of the species. The same is true for the world's economies. Some systems do better than others, and the relative performance of economies alters over time. If one economy fails, world institutions, such as the IMF, can offer support to get it running again. If the world economy is integrated into one common system based on common rules, there may be short-term advantages, but any slight flaw could bring down the whole system. I first published this perspective on economic interdependence prior to the financial crash of 2008 (McManners 2008: 171–2). I was not to know then how prophetic it would prove to be. The speed of the financial crash astounded everyone, but people were looking for specific causes. Few people realized that the prime cause is the global system itself. Occasional total system collapse is a feature of one single interconnected system. Economies go through cycles: if the economy is global then we can all share in the upside, but we all also descend together. Desynchronized national economies bring macro stability to the world economy. The price of this stability is a weaker growth phase, but I suggest that this is a trade-off worth making.

Through the course of 2009, world policy makers looked to strengthen global oversight of markets. For the people close to it, this seemed to be the way to rescue the current system. The most effective action required is quite the opposite. The application of the principal of subsidiarity would reverse integration of the world economies and bring back variety and independence. This stabilized world economy would then go through local crisis as mistakes are made, but such crises could be contained and the whole system would remain stable and secure. See Chapter 9 for a fuller discussion of specific measures to bring finance back under the control of national governments.

Towards a Global Commitment to Subsidiarity

There is always a danger that the principle of subsidiarity is forgotten or ignored when we observe what we regard as poor management or weak policy in other countries. The only reason to influence change is if the effects spill over to have detrimental effects on other countries or the global ecosystem. Attempting to be dictatorial at world level, even with very good intentions, is counterproductive. Not only is there is no need to meddle in the internal affairs of another country, according to the principle of subsidiarity it is important that we don't.

Efforts to deal with global challenges should concentrate on global targets and outcomes. For implementation, we must enforce the principle of subsidiarity, for two reasons. First, human society responds more intelligently and effectively at small scale. Second, world society needs a variety of societies and economic systems to ensure macro stability.

There is one significant potential contradiction that is exposed through a focus on subsidiarity. At global level, we must avoid putting national self-interest first. At the same time as encouraging less interference by global bodies, global governance has to be reinforced to ensure that the global agreements truly reflect global needs. There will always be the temptation to use global agreements to pursue purely national self-interest. This is contrary to good global governance and has to be resisted.

In this chapter, I champion a focus on local decision making and variety in the solutions we consider. Subsidiarity does not mean that all decisions are local, of course. There needs to be a hierarchy of higher levels of coordination and authority. At each level, a sustainable balance is set and then enforced. The prime level, at which the most important compromises are made, is the nation state. Governments have responsibility across all aspects of society and have the prime role in coordinating, leading, responding to the wishes of the electorate and forcing compliance.

4

The Primacy of the State

The government of the nation state is where primary responsibility resides for human society. Each country has a unique combination of culture, geography and population requiring an appropriate policy framework. Differences are inevitable. This allows us to take pleasure in the breadth and variety of human culture, but, more importantly, variety ensures macro political and economic stability. Differences facilitate adaptation at the global level. Instead of global hegemony, the world needs a rich variety of cultures, societies and economies.

The nation state is singled out for particular attention because this is the level of world administration with the most capability to set sustainable policy. This is where the combined value of hard and soft power is greatest. Governments have power over borders, legislation and fundamental areas of policy and can call on a sense of nationhood, loyalty and pride. They, therefore, have the most influence over building sustainable society.

I use the term nation state to convey the concept of a country in which the legal structures coincide with people's sense of nationhood. In an ideal world, all countries would be nation states. In the real world, the legal entity, that is the state, may not have the allegiance of all segments of the population; hence the soft power of the state may be less than I assume. There are also states that are too small to wield hard power across a broad range of policy areas. An example is Luxembourg, which is reliant on surrounding countries for much of its support infrastructure and policy framework. Despite these difficulties, the concept of nation state is useful in building a conceptual framework.

People are born into a family, a culture and a nationality. Nationality brings with it rights of residence and access to the support of the state. It also brings responsibilities such as national service. We have no control over which parents we have or where we are born. As adults, we can change our nationality, but this is not a common occurrence. The inertia of society encourages us to be loyal to our roots.

Through a person's early formative years in education and growing up within society, they are exposed to the culture of their birth. Open young minds are easily influenced to take certain values to heart. As adults, these values become central to who we are.

This process of indoctrination is so obvious that we do not think about it. Civilized society depends on values passed down the generations. Without deep-seated values, civilization cannot exist. This bedrock of culture and values is the foundation of human society.

One of the values we need to instill in the next generation is adoption of the concept of sustainability. This is very different to the ruling culture in most countries over the last half century, with a few notable exceptions. The Himalayan country of Bhutan brought in the concept of Gross National Happiness (GNH) to guide building an economy that retained Bhutan's culture and spiritual values as the country opened up to modernization following the death of King Jigme Dorji Wangchuck in 1972. People born in Bhutan are brought up to understand what living a happy life in tune with their surroundings entails. The lessons they learn are appropriate to the locality and the particular circumstances of the country.

Such isolated beacons of sustainable self-determination need to be admired and supported. This is rare. The Western model of progress is presented as the one to adopt. It was not long ago that the World Bank and other development agencies sought to reform 'backward' nations that conformed to other models.

People will always be able to observe other countries and cultures and copy what they like and admire. The high-consumption model is attractive when seen from afar. Those of us who live in the West can see that we do not live in perfect societies. When people are drawn to join our societies, they are often disappointed by the reality when they arrive. The West needs to be much more aware of the limitations of Western societies and show more humility in understanding that there are other ways to run society. At this vital turning point in human history, we must admit that we have made mistakes.

I sought to persuade an audience at the London School of Economics in 2007 that we in the West should admit to our mistakes with regard to city design. The theme of the conference was the challenges of cities in international development policy. In my paper, *Cities for People: Removing Cars from Urban Life* (McManners 2007), I argued that Western cities have grown around the needs of the car over the last half century and that it was time to return cities to people. I went on to argue that our advice to the developing world should be to avoid repeating our mistakes and bounce ahead to the sort of cities that the West should now be aiming for. As a resident of the West, there was a danger that I would be seen as a hypocrite by seeking to deny many people in the developing world the aspiration of car ownership. Interestingly, a number of representatives of poorer and less-developed countries supported my view. Opposition to my view was led by World Bank officials (Westerners), who argued that the business case for roads is stronger than for other public transport options. These officials, brought up in a society dominated by the car, do not have the experience of alternative city design. They are setting policy to roll out 'car-centric' cities around the world whilst I and others are championing the adoption of a different model.

The imposition of unsustainable policies on other countries, based on the experience of Western development over the last half century, will be seen in hindsight as incompetent. The sad fact is that the guilty people are acting with the best of intentions. They want other countries to enjoy similar levels of success to their own home countries. We must move quickly to change the mindset of the West, not only to transform our societies, but also so that officials drawn from our culture are equipped to offer useful advice to developing nations.

Once the transformation of Western society is under way, we will be able to parade and champion a different economic model. If we procrastinate, and any changes are perceived as slow acceptance of unpleasant medicine, other countries may begrudge our material success and will not want to follow our new example. Poorer countries may insist on reaching Western levels of consumption before changing course. It is important that we embark quickly on an enthusiastic improvement in our society that other countries will envy and copy.

Reinforcing the Nation State

The worst human suffering is to be found in failed states, where the apparatus of government has collapsed. Without a functioning government, criminals and people with evil intent have free rein. In such a lawless dystopia, the environment has even less chance of protection than the people. Whatever the type of government, any government is better than no government.

In an increasingly globalized world, one worrying possibility is that, as we reach the limits of the Earth's resources to support a growing population, the collapse of one state could trigger a domino effect with whole regions descending into anarchy. If world society does not take a different direction in the coming decades, the chances of this calamitous outcome shift from an outside possibility to a near certainty (McManners 2009). States need to put a higher priority on protecting access to resources and becoming self-sufficient for key functions. Those who live in stable countries with competent governments tend to take it for granted that the state will always be there. The integrity of the state should not be taken for granted. We must work hard to protect and reinforce it.

In large countries with a wide geographical extent, it is easier to balance resources with demand, but it is harder to implement a cohesive nation state. The former Yugoslavia is an example. Following the Second World War, the country united under the leadership of Marshall Tito, who had been the leader of the Yugoslav resistance movement during the war. As Prime Minister, and then President, he provided a shared vision of nationhood until his death in 1980. After his death, his unifying vision faded and it was only a matter of time before each of the nations in Yugoslavia started to exert its national identity. After the wars of the 1990s, stability has again returned. The nations of Serbia, Croatia, Slovenia, Macedonia and Bosnia are back in control of their own affairs. Tranquility is slow to return as these new states learn to deal fairly with their minorities, but, from a regional perspective, the new structure is likely to be more stable if people feel content that they have regained control over their affairs.

Large countries will always have difficulties because it is so hard to build and retain a common identity. The solution is a federal structure in which regions enjoy high levels of autonomy, thereby engaging and empowering the population within the political process. There are many successful examples, such as the United States and Germany, which combine advantages of scale at the country level with the effectiveness of empowerment at the state level. I contend that one reason for the stability and economic success of Switzerland and Canada is that they are among the world's most decentralized federations.

The EU is an interesting example. Formed as a regional trading block in 1957, it has expanded from its original six members to 27 members in 2009. The EU has been attempting to set an example since sustainable development was included as an overarching objective of EU policies in the Treaty of Amsterdam of 1997 (Bär & Kraemer 1998). Taking these policies forward into a stable and sustainable European society will be hard.

Within the EU, discussions continue about deeper and wider integration. Many pro-European policy makers are hoping that this will lead to a federal European state. European leaders have to be careful in shaping the future political landscape not to overextend the reach of its institutions. EU directives and EU borders are specific and well defined. The concept of a nation of Europe is much more diffuse and will probably never be accepted. It would be dangerous for the EU to seek to take more control away from its members than their people desire or will accept. It would also be risky to draw in new countries that are not committed to the European ideals and values unless the relationship is a loose, less intrusive collaboration.

Over the last half century, the ruling economic orthodoxy has been a commitment to free trade. This has led to the opening of borders and the establishment of a number of regional trading blocs such as the EU, the North American Free Trade Agreement (NAFTA) and the Association of Southeast Asian Nations (ASEAN). The economic benefits have flowed. Policy makers have been much slower to understand the negative effects that free trade brings. Opponents of free trade often campaign against necessary and sensible economic reform, and so are ignored. But this obscures the genuine argument that free trade can undermine countries trying to set a sustainable policy framework that balances local social and environmental circumstances.

Reinforcing the integrity of the nation state is vital to building sustainable societies, and economics has to fall into place in support of this aim. Free trade and open markets, unless part of a much more complex policy package, are no longer appropriate as the focus of economic policy.

Strength in Variety

There are many situations where strength derives from variety. This applies to the ecosystem, society and the economy. At world level, a collection of different, loosely coupled economies will be much more resilient than one interconnected world economy.

The global economic boom of the last few decades coincided with the widespread adoption of the policies of the Washington Consensus.[1] This was not just a coincidence – the Washington Consensus was a good policy framework for delivering short-term economic outcomes. Such policy gives scope and freedom to squeeze the maximum financial return out of world society.

For a number of years, I have been questioning the assumption that free trade and open deregulated markets always constitute good policy. In workshops with MBA students, I would ask, 'Is globalization reversible?' The debate that followed would not last long. There was little appetite for such a discussion. When the financial figures were looking so good, there was an unwillingness to question what lay behind them.

At the time of writing, in the midst of the recession of 2009, the circumstances exist for a genuine debate over the possibility of reversing globalization. But progress is slow. Pro-globalization philosophy has become so ingrained that, in response to the current crisis, discussion over world economics has focused on greater conformity, increased regulation and a strong drive to persuade governments not to focus on national solutions, citing protectionism as an evil to avoid. In effect, world leaders are looking for more globalization, not less.

The lesson we drew from our experience of the last few decades was that closer linkages between national economies reduce the risk of economic collapse. We observed this to be true on a country-by-country basis. We did not understand that economic globalization increases the risk of a systematic failure that ripples through the whole world economy. In 2009, we now know this to be the case, but find it hard to accept that the basis of policy has to change.

It is evident that a single global economic system is not feasible or desirable. It is less widely understood that a collection of separate economies that are clones of one economic model should also be avoided, because the particular attributes of this one model would be greatly amplified. For example, the US

1 The term 'Washington Consensus' was coined by John Williamson (1990) to describe a set of ten specific economic policy prescriptions constituting a 'standard' economic reform package including trade liberalization, liberalization of inward foreign direct investment, deregulation and privatization of state enterprises.

economic model, if rolled out worldwide, would destroy the ecosystems of the planet. This is not intended to be anti-American; it is a simple statement of fact derived from logical analysis. Other economic systems, such as the policies of Bhutan, may be excellent at preserving the ecosystem but, if adopted universally, might never afford humankind the capability to, for example, migrate from this planet to explore and colonize other worlds and other solar systems.

The governments of some countries will be better than others at delivering improvements to the lives of their inhabitants. These countries will be the places to look for policies to copy and adapt. Other governments will provide practical demonstrations of policies to avoid.

When analyzing the success of any particular society, economic policy is just one consideration. Copying an economic model, without understanding the context, may result in learning the wrong lesson. An example is the policy of privatization that was implemented in many countries with the advice of the IMF and World Bank. Experience of the policy in the UK, where it was implemented by the Prime Minister, Margaret Thatcher, in the 1980s, has delivered what was expected. The process of transferring ownership has been transparent and fair. The privatized industries are regulated by independent experts appointed by the government. Privatization in other countries has not worked so well. In many African countries, friends of the ruling elite have been the main beneficiaries. In Russia, the privatizations of the Yeltsin years left oligarchs in control.

One response by external advisors is to cite corruption as the cause of such policy failures, but this is going further down the road of imposing our views on the internal affairs of another country. Real progress will come from looking for a solution to the delivery of public services that is appropriate to the area concerned, be it Africa, Russia or anywhere else. At global level, a minimum of economic policy is required, and any policy we adopt has to be different to the policy of the last few decades. As world leaders tackle the deepest recession for over 60 years, there is widespread support for change. It is harder to agree the shape of the new economic policy package. I argue that one aspect of the new system must be explicit policy objectives aimed at retaining a variety of independent economic systems.

Achieving the objective of variety in economic systems will not be hard. Even though economic systems with a good track record are likely to be followed widely, countries have a natural tendency to forge their own way. To ensure that one economic policy framework does not monopolize the world economy, all that is required is to stop forcing conformity through the advice offered by world institutions. This will be enough to allow economic ingenuity to flourish as people's natural tendency to find their own solutions prevails.

Sustainable Nation States and Regions

A sustainable world requires that we have sustainable nation states and regions. World institutions will never have the authority or support to impose a global sustainable solution. It falls to the nation states to act, alone or through regional arrangements. It is worth noting that I do not argue for reversing our technical progress or for localizing all our activities. Sustainable policy requires the application of advanced technology and trade between nations, but its nature is different. Technology should be used, not for new gizmos, but to deliver sustainable solutions. Trade should alter to match real needs rather than to arbitrage between the competitiveness of one economy versus another.

Sustainable living should be easy, provided we adopt appropriate structures for society within which people can negotiate a balance between social need and available sustainable resources. The principle of subsidiarity pushes decisions down to communities. These communities operate within the nation state, where prime decisions are made. Under the stewardship of the state, we can build a sustainable world. Each state should be free to choose its own way and set fiscal policy to suit.

5

Green Economics

This chapter considers how government fiscal policy and the use of market mechanisms can support a sustainable society. Specific policies are cited, but by way of example rather than as a blueprint for government policy. As the discipline of 'green economics' described by Cato (2009) gains credibility we should expect 'distinctive new solutions' to emerge (Kennet and Heinemann 2006), influenced by Anderson (2006), Reardon (2007) and others. The New Economics Foundation, an independent think tank based in the UK, is at the forefront of attempts to define 'the new economics'. A powerful critique of the contradictions in conventional economics has been developed (Boyle and Simms 2009) but a coherent and workable alternative economic paradigm has yet to evolve.

In this chapter I do not attempt to define green economics or a new economic system but to focus on achieving the following three outcomes:

- make carbon markets effective;
- appropriate localization of supply chains;
- strike a balance between human needs and nature.

In a sustainable society, fiscal policy should encourage behaviour that is sustainable and penalize actions that are not. Government taxation policy will shift towards taxing resource inputs and environmental impacts and away from taxing income. The final end state will be very different to the tax systems of today. The outcome is hard to predict in detail; each policy change will need to be tested and unforeseen problems will need to be ironed out.

The more I have examined a sustainable society, the more I am drawn to the benefits of increased government intervention. Regulation and taxation seem to provide more direct control than markets. However, this relies on an intelligent and all-powerful benevolent state. The practical evidence points towards markets as being rather better at encouraging human ingenuity to flourish. But we should not have blind faith that the market is always right. Markets work best when the solution is not obvious and a range of competing options play off against each other until the market settles on a winning solution. When it is clear that certain costs must be higher – to force the transition to sustainable methods – taxation works better as a means of providing unambiguous, firm figures on which to plan investment. As sustainable policy is better understood, and we have worked out how a sustainable society should operate, then carefully crafted taxation will be important. In the formative stages of the Sustainable Revolution, policy will be in a state of flux as new ideas and new technologies are developed and tested. This situation requires innovation and ingenuity, and is where the flexibility and openness of the market can excel.

The regulation of markets at the national level is generally better than regulation at global level. At the micro level, market forces are automatically constrained by community oversight. 'Bad' behaviours are noticed and action is taken. At this level, self-regulation can work. At national level, markets need more formal oversight by governments, which have the power to regulate as necessary. At international level, markets are subject to little effective oversight and enforcement is weak. As a general principle, it would be unwise to trust international markets to deliver sustainable outcomes.

In recent years, there have been developments in world society showing constructive intent, for example the growth of carbon trading. This has been at the heart of policy makers' discussions over a response to climate change. The European Union Emission Trading System (EU ETS) set up in 2005 is the first sizeable market for carbon. In 2009, the United States made the first tentative move towards introducing carbon trading legislation with the Waxman-Markey bill (US Congress 2009). It is vital that this momentum continues, but few people have taken the time and effort to look beyond the theory to examine and anticipate the long-term effects in the real world. Policy makers will discover by experiment what works and what doesn't, but they will need to learn quickly. Later in this chapter, I provide some pointers to short-circuit this learning process.

My focus is on the medium-term evolution of green taxation and market economics. I have come to believe that in the long term very different systems of taxation are needed. However, the risks of campaigning for total change at this early stage are too great. Some advocates of green economics may be disappointed that I do not go far enough. This must be balanced with the needs of policy makers in the real world, who require guidance on what is achievable and can be sold to a sceptical public. To put my specific proposals in context, first it is necessary to discuss some core concepts.

Sustainable Taxation

Taxation has the evident purpose of raising government income, but it also has a role in influencing behaviour. In a sustainable society, taxation will be designed to achieve economic, environmental and social outcomes.

Economic outcomes

The economic purpose of taxation is to raise the income a government needs to fund its expenditure. This leads to identifying transactions and cash streams that can be taxed in such a way that avoidance is difficult and collection is administratively easy and politically acceptable. This direct economic purpose – to raise government income – has shaped modern tax systems. A number of taxes are well established and have robust systems of collection, including income tax, sales tax (Value Added Tax, VAT), corporation tax, capital gains tax and inheritance tax.

In setting tax rates, politicians have to raise the money needed to balance the budget in a package that appears fair and equitable. The focus is currently on drawing tax income out of society with the least political pain. This focus will change as the benefits of environmental and social purposes of taxation are better understood and more widely applied.

Environmental outcomes

Environmental taxes, such as carbon taxes and taxes on waste, are applied with the prime purpose of changing behaviour to protect the environment. In a sustainable society, these taxes will increase in importance. In simple terms, they will involve taxing the 'bad' and supporting the 'good'. For example, taxing pollution generates income that can be used to support adaptation away from dirty processes. Often the income is balanced against related expenditure, so that the overall cost to the government is neutral. Another method is to use environmental taxes to increase the overall tax take. This can have a double benefit in raising income and changing behaviour. However, it can also lead to a perverse incentive to continue the 'bad' activity in order to maintain government revenue.

Social outcomes

Using taxation to deliver social outcomes is well known with regard to health, for example, the taxation of cigarettes. A paternalistic and authoritarian government could simply ban cigarettes, but a more democratic approach is to tax them heavily at a rate that exceeds the related expenditure on the treatment of diseases caused by smoking. Another obvious social role for taxation is in redistributing wealth by taxing the rich more than the poor. There are less obvious measures where taxation can help to deliver social outcomes. In a world that is increasingly automated, governments will have to consider how to engage people in work. In theory, a government could tax automation in order to support the social outcome of more jobs – though this would be an odd tax to introduce. The same effect could be achieved by reducing or even removing income tax (as work is considered a 'good') and replacing the lost revenue by taxes on material inputs and other core green taxes. (This theme is returned to in Chapter 9, p. 124.)

Core Green Taxes

It is worth examining how taxation would be structured if starting from a clean sheet with the intention of supporting a sustainable society. Raising government income in a way that discourages the use of natural resources would then be at the core of sustainable taxation. The most valuable, finite and universal resource is land, and the use to which it is put it is vital to the health of society and the ecosystem. It would make sense to make land a target of core green taxation. This idea has been around since Henry George championed the concept (1879). He argued that all land should be made common property and users charged a rent payable to the state. Modern proposals do not go so far as to confiscate all private land but land tax[1] makes landowners pay a full rent to – in effect – be allowed to continue owning the land. This tax is different to normal property taxes in that it is applied to the rental value of the land, not the value of the buildings erected on the land. Such tax is secured against title of the land so it is easy to collect, and, if unpaid, the land can be taken in payment of the debt.

1 Land tax, site rental tax and land value tax are synonymous terms.

The argument for land tax is based on economic efficiency and fairness. It discourages landowners from holding on to land merely for speculation, thus improving the availability of land and its efficient use. Empty sites are brought into productive use. In terms of fairness, the argument is that ownership of land is concentrated amongst a small group of people and that land tax is the only way to break the monopoly (Porritt 1994). Tony Vickers (2007) makes the case for a land value tax in Britain based on both fairness and efficiency, arguing that this would be part of shifting the burden of taxation from enterprise to resource usage.

The 'greening' of the taxation system should include shifting the tax burden from income tax to land tax, but this needs to be introduced with care. There is not only a political difficulty but also a fundamental problem. Land tax, in its simplest form, could perpetuate the process of exploiting land at the expense of the ecosystem. The closing section of this chapter – 'Urban Eco-Balance Tax' – addresses this weakness.

Core green taxation could also focus on other resource inputs, ranging from fuel and commodities to timber and water. The quantity and pace of exploitation can be slowed by having resource inputs as the focus of the tax system. The economic incentive then supports the design of processes that minimize such inputs and maximize recycling and reuse. Where resources can be extracted sustainably, resource input taxes can be part of core taxation, both to generate income and hold consumption in check. Currently, fossil fuel is a prime target for taxation, and will remain so for many years, but basing core tax income on such an unsustainable resource is itself not sustainable. As fossil fuel is squeezed out of the economy, tax receipts will drop. This is when transition taxes are more appropriate.

Transition Taxes

Transition taxes are designed to force change in society and are usually of limited duration. The idea is to tax unsustainable activities to reduce them to a sustainable level, or to eliminate them entirely. Economic choices are then steered towards more sustainable methods or processes.

The effect of the tax can be amplified by earmarking the revenue to pay for corresponding infrastructure investment. For example, taxing fossil fuel reduces consumption through the coarse but effective mechanism of price. The taxes can be invested in building infrastructure that is less reliant on fossil fuel, such as improved public transport. The tax revenue can also be used to prevent poor families from slipping into fuel poverty through improvements – such as better insulation – to reduce bills through less consumption (rather than cash subsidies). Such earmarking of revenue also makes it possible, politically, to drive taxes higher sooner. The population is more likely to support a system that is seen to be fair when the receipts are channelled into helping society to adapt.

As a transition tax achieves its purpose – stamping out unsustainable practices – the tax receipts drop to zero. Therefore, transition tax revenue should not be used to fund general government expenditure lest a perverse incentive arises to continue with the activity to bolster government finances.

In some cases, the tax would no longer be required after the transition period if the end target were an outright ban. The date of the ban could be announced well in advance, with the transition tax being levied at an escalating rate according to a published schedule. This would give business, government departments and consumers firm figures and a timescale to support investment plans. An example of this approach could be with regard to categories of waste. Escalating taxes could provide a strong disincentive leading to the total elimination of certain types of waste.

Transition taxes are simple to apply and are the preferred choice in situations where the parameters are easy to define, as firm figures can be produced to support investment decisions and business plans. In complex situations where it is not clear what the most cost-effective solution is, market mechanisms may work better than transition taxes.

Market Mechanisms

Market economics has played an important role in developing advanced modern society and it can also be used to help shape a modern green society. The reputation of free markets has been tarnished by unintended negative consequences, such as environmental damage caused by overexploitation of natural resources. This should not obscure the market's record of success in allocating resources efficiently. Market mechanisms can also be used to reduce waste and pollution. The particular challenge in the coming decades is to curb carbon dioxide emissions, and carbon markets have been established for this purpose, but there is a huge challenge to make such markets effective in delivering the intended environmental outcomes.

In addition to potential economic efficiencies, markets can also have political advantages. When the oil market climbed steeply to over $147 a barrel in July 2008, this was due to the market. People complained, but accepted that market forces were to blame. If politicians had forced the price of oil to that level through taxation, there would have been an electoral backlash. The market can insulate politicians from tough decisions on setting tax rates.

This section focuses on markets because of the complex and interconnected nature of the transformation that is required for society to become green and sustainable. Policy changes will be necessary in areas that overlap and interact. Detailed top-down planning will struggle to cope, no matter how clever and complex the plan. The invisible hand of the market can handle such fluid complexity to find the most cost-effective solutions. The key factor with regard to markets is that they are only as good as the rules and regulations that hold them to account. Markets should be designed carefully, and appropriate regulation applied, in order to achieve the desired outcome.

The weakness of global market mechanisms

Where a market is allowed to extend beyond one state to become a world market, the problem of governance arises. This applies to world trade and financial markets as discussed further in Chapters 7 and 9, but it also applies to any market.

Good governance of a global market needs an international organization under the auspices of a body such as the UN. Such organizations find it hard to implement policies that match the broad responsibilities of governments. Part of the problem is the difficulty of top-down control in a complex world. Another part is the reluctance of countries to cede real power to international agencies. Global governance is therefore weak and likely to remain so. Markets are best controlled at national level, where governments have the authority to implement effective and balanced controls to ensure the market delivers benefits to the society as a whole. This principle applies to all markets, but particularly to markets for pollution where the strongest countries can offload pollution (or dirty processes) on to countries with weaker regulation and less ability to cope.

Markets for pollution

Regulation is required in order to tackle pollution, but markets can often leverage much greater effectiveness in driving change to deliver pollution reductions. The system of cap-and-trade creates tangible financial rewards for environmental performance by turning pollution reductions into marketable assets. Once pollution reductions have a value, the system prompts technological and process innovations that reduce pollution down to or beyond required levels.

The market mechanisms can work in a number of ways. A key aspect is the setting, and enforcement, of a cap. The cap on total emissions in the market is set with advice from experts and then permits-to-pollute are traded in the market. One way to initialize the market is to hand out permits based on current emissions. Each year, the allocation can be reduced proportionally until total emissions are lower than what is judged to be an acceptable level. The market participants can either make reductions to operate within their allocation, or buy permits to continue emitting pollution. Each participant calculates the cost to them of pollution reduction and compares this with the market price of permits. In this way, investment flows to where the most cost-effective pollution reductions are to be found. Companies that find they can make reductions easily can profit by reducing emissions below their allocation and then selling their permits in the market.

The reduction of sulphur dioxide (SO_2) emissions in the United States is a good example of the effective use of such a market. The 1990 Clean Air Act established a market system to reduce the levels of pollutants causing acid rain. It was targeted initially at SO_2 and was later extended to include nitrogen oxides. A decreasing cap on total SO_2 emissions was set, which aimed to reduce overall emissions to 50 per cent of 1980 levels by 2010.[2] By 2007, SO_2 emissions were below the long-term emission cap three years before the statutory deadline. Burtraw (2000) concluded that, 'Emission allowance trading ... has contributed to significant cost reductions, compared with original forecasts of cost.' The US Environmental Protection Agency (EPA) (2009a) estimated that public health benefits from the emission reductions exceeded programme costs by a margin of more than 40:1. The flexibility of the market supported a range of strategies including using lower sulphur coal, switching to gas, installing scrubbers and

---

2 SO_2 emissions in 1980 were 18.9 million tons; the cap set for 2010 is 8.95 million tons. SO_2 emissions in 2007 were 8.9 million tons.

retiring older plant. This flexibility of implementation is where markets for pollution allowances can excel.

The 1990 Clean Air Act is a federal law covering the entire United States. Interestingly, the law recognizes that states should lead in implementing the Clean Air Act, because pollution control problems often require special understanding of local industries and geography (EPA 2009b). This is a further reminder that localized decision making in policy implementation is more likely to lead to balanced sustainable solutions. Another interesting development has been the emergence of not-for-profit environmental groups, such as the Acid Rain Retirement Fund (ARRF), which have been purchasing pollution allowances in order to take them permanently out of the system. The ARRF owns the legal right to emit 122 tons of SO_2 per year – a right that it will not be exercising (ARRF 2009). This leads to additional emission reductions beyond the target set by the legislation.

The markets can become quite complex and include a futures market to allow hedging of allowances and to give more certainty to investment decisions. One obvious enhancement is to auction the permits. This avoids the evident unfairness that the biggest polluters, who have presumably done least to make improvements, are handed the biggest allowances. An auction provides a level base for all market participants – and considerable income for government. It can be argued that the proceeds should be earmarked for related expenditure such as funding for research into cleaner processes or aid to developing countries to help them implement reductions.

Making Carbon Markets Effective

Carbon dioxide emissions from fossil fuels are the prime cause of climate change and must be controlled. Carbon trading has potential to help society to make the transition to other energy sources.

The EU ETS is the largest carbon market so far. It has been a good test bed for carbon-market mechanisms. The first lesson has been that the free allocation of permits, based on past emissions, handed a windfall profit to some of Europe's biggest emitters. A proportion of future allocations will, therefore, be by auction. The second lesson has been that over-generous allocation of permits undermined the price of carbon, which dropped to below €10 per tonne in spring 2009. The EU commission will have to be much more robust in setting the overall cap in the face of resistance from national governments seeking to win big allowances for their own industry.

There are also Europe-wide concerns that certain industries will find it hard to compete in the world market due to inclusion in the EU ETS. One of these is cement manufacturing, which is a large CO_2 emitter. Aviation is another.

In July 2008, the European parliament voted to include all flights in the EU ETS from 2012, including flights from outside Europe. The proposal was criticized for seeking to reduce CO_2 emissions by only 3 per cent compared with historic emissions over the period 2004–06 and for putting just 15 per cent of permits out to auction (EU 2008b). The reality is that both the cement industry and aviation need to be completely reconfigured to fit within a sustainable economy; the sooner the invisible hand of the market is allowed to pull the lever of much higher energy costs, the sooner the transformation can take place.

Open global markets are making politicians wary of taking the necessary action. They should adopt the policy framework of proximization in order to be able to implement sustainable policy. Governments will then have greater freedom to protect local industry from foreign competitors that are not doing enough to squeeze carbon out of their economy.

The immediate effect of a carbon market, such as the EU ETS, is to shift consumption to less carbon-intensive fuels such as gas and oil, and away from coal. This short-term improvement masks long-term problems building up in the carbon markets as cleaner fossil fuels are depleted.

At the time of writing, the main problem is that, at world level, there is no binding cap on CO_2 emissions. It is doubtful that the world community will be able to agree and then enforce such a cap. Even if a global cap were to be agreed, politicians will be tempted to back off if the economic pain becomes too great. It is hard to envisage a tightly regulated world carbon market, and dangerous to rely on the assumption that one could be established. Negotiations could drag on for decades with a whole series of movable targets accompanied by regular reports documenting failure to meet them.

A carbon market without a robust cap (that is reduced year-on-year in line with scientific advice) will fail to reduce overall CO_2 emissions. Instead, the market will ingrain reliance on fossil fuel into the world economy in the belief that the market will find a solution over time. The 'benefit' of the carbon market will be to smooth the transition from clean fossil fuel to dirtier fuel, such as coal and oil extracted from oil sands or oil shale. As the clean fuels run down, the market will shift smoothly to lower-grade fuels that deliver less net energy for the CO_2 emitted. For example, coal combustion emits almost twice as much carbon dioxide per unit of energy than the combustion of natural gas (Hong and Slatick 1994). As we struggle to impose a cap on global emissions and shift from natural gas to coal, we will have to halve energy usage simply to maintain a standstill in CO_2 emissions.

Another problem is that aspects of the world market, such as the Clean Development Mechanism (CDM), are less effective than had been hoped. The CDM, set up under the Kyoto Protocol, allows countries to purchase carbon credits for the implementation of low-carbon projects in developing countries. In theory, any reduction anywhere in the world will contribute to reduced levels of CO_2 in the shared global atmosphere. One difficulty is certifying that the projects are additional to what would have happened without the award of carbon credits. An even more fundamental problem is that without a capped and tightly regulated global market, such international transfer of carbon credits will not achieve overall reductions. CDM-funded projects have provided some useful development aid, but, in a leaky global carbon market, the overall effect of the purchase of carbon credits does little more than provide justification for delay in making reductions by the purchasers.

If developed nations are serious about forcing the pace of change towards a zero-carbon society, carbon markets must be kept national, or perhaps regional, such as within the EU. Even though making reductions in developed countries will be more expensive, this investment will be leveraged many times over through the example it sets. As richer societies are transformed, the advances in know-how and technology can be used to prove and parade a low-carbon model. The West will then be able to offer credible advice. As the developing nations see that serious action is being taken, they will be amenable to the idea of bouncing forward to the sort of twenty-first century societies championed in this book. Over the long term, this will deliver far greater carbon reductions than making marginal reductions on a project-by-project basis that is paid for by delaying making reductions in the developed countries.

National carbon markets can work in tandem with taxation to deliver real reductions, providing the delusion that a world carbon market can solve the problem of CO_2 emissions is avoided. Relatively clean fossil fuels, such as gas, are available to tide us over the transition to renewable energy sources. This opportunity must be exploited by national carbon markets with a fixed cap, fixed timescale and the clear final outcome of a zero-fossil carbon economy.[3] A floor should be set to the market to ensure that the case for investment in renewable alternatives is sound. This could be achieved through direct taxation on fossil fuels at levels as high as politically acceptable. The market could then be made responsible for determining price increases above this level to keep within the cap. This is the realistic and effective way that carbon markets can be tightened to deliver on their potential as an economic tool to reduce society's reliance of fossil fuels.

Appropriate Localizing of Supply Chains

There are multiple reasons why localizing or shortening supply chains have green economic benefits, ranging from reducing transportation costs to connecting the population with its resource base. In a sustainable society, drawing on resources from the locality to allow a sustainable balance to emerge, should be the normal default option. For example, short supply chains in agriculture leads to more transparent processes, healthier food and greater security of supply. Another example is the sale and consumption of many manufactured products.[4] Within a localized system, true recycling is possible, using cradle-to-cradle manufacturing processes and through developing a close relationship between manufacturer and consumer (McDonough and Braungart 2002). To achieve shorter supply chains, the simple economic lever of increased transportation costs can be used, but a more specific economic lever would be useful, particularly during the transition phase.

One method is a distance-to-market tax (McManners 2008: 138–9) that would be levied on the number of transport miles used to bring the product to the point of sale. However, there are a number of complexities to calculating a comprehensive figure for the transport miles associated with a particular product. My proposal is for a simple tax that assumes a single place of production. As we succeed in localizing many more consumption cycles, this assumption will apply rather better.

3 The target of a zero-fossil carbon economy seems like an impossible ambition, but without such an unequivocal end-point to aim at near- and medium-term action will fall short of what is required.

4 Most products can be made locally if greater emphasis is placed on sustainability in contrast to a narrow focus on economies of scale. There will remain highly specialized items and equipment, such as advanced aircraft, for which the most sustainable solution is centralized production in global or regional factories.

Reducing food miles is an obvious target for such a tax. Food used to be a local business supplemented by exotic produce from further afield. Now, a quarter of all heavy-goods vehicle traffic in the UK is for the transportation of food (Smith et al. 2005). The huge increase in food miles is a fairly recent aberration in our economic system which has been encouraged by very cheap transport costs and the dominance of a small number of large food retailers pushing for economies of scale. If a distance-to-market tax were implemented, local and seasonal produce would be relatively cheaper, whilst exotic foods from remote corners of the world would be more expensive. The tax would hit people who could afford to pay; poorer people could keep their shopping bills low through buying local produce. The concept of a distance-to-market tax could apply, not just to food, but to all items sold in order to strengthen the economic case for true recycling.

The basis of a distance-to-market tax would be a location code included in the product code. At the point of sale, this code would be compared with the location of the sales outlet. Tax could be set per unit of distance as a very small percentage. Governments would probably wish to tinker with different rates for different categories of items (as they do now for VAT). Items assembled from a number of components from different suppliers would be complex to deal with. But, over time, we should be aiming for a system of manufacturing where complex products are produced by clusters of co-located companies in which total lifecycle design is the norm and local recycling automatic.

Until recently, such a distance-to-market tax would have been a huge bureaucratic overhead. Now, IT systems used by retailers are highly advanced. It would be a relatively simple task to add in a few lines of code to calculate and collect the tax (provided the product coding system includes location of production). There would be arguments over the details, of course, as with any new proposal. But it would be feasible to introduce such a tax to speed the process of shortening supply chains.

A distance-to-market tax may not be universally applicable. For example, commodities such as fuel supplies will be treated differently. A green society will, as far as possible, use renewable local supplies, but each location is different. For example, locations towards the equator have much more sunshine and much more capacity to generate renewable power. Far north and far south locations will need to supplement local renewable energy supplies by sustainable global energy flows. The key issue here is a guarantee of the sustainability of the source. Biofuel is an example. Countries that commit to the principle of sustainability should not import biofuel produced on land that is cleared rainforest or – in a world short of food – on agricultural land that could be used to grow food crops. The really useful biofuels are the third-generation biofuels[5] under development, which could be produced in otherwise unproductive desert regions. It would probably make sense not to apply a distance-to-market tax on truly sustainable fuel supplies.

Other exemptions to a distance-to-market tax will include encouraging a certain amount of sustainable trade to help poor countries, in products that can be transported efficiently, for example by ship. It would clearly be important not to exempt fresh produce delivered to our markets by air freight. UK consumers spend over £1 million each day on fresh produce, including fruit, vegetables and flowers, which is air freighted from Kenya, Ethiopia, Tanzania, Zambia and Zimbabwe (IIED 2007). Some people argue that such unsustainable food transport arrangements should be banned. Others defend the trade, citing the economic benefits for the farmers, despite the fact that a large proportion comes from farms owned and leased by the importers. I prefer to use the distance-to-market tax and let the market decide. What would emerge is a very low-volume trade in expensive premium produce.

In addition to the immediate challenges of migrating society away from fossil fuels and localizing supply chains, we also need to stabilize land use. Striking a balance between human demands for land and nature's needs is another example of applying the concepts of green fiscal policy.

Urban Eco-Balance Tax

The general concept of land tax as part of core green taxation was discussed earlier in this chapter. In this section, a particular variant of land tax, which I term urban eco-balance tax, is considered in order to tackle the fundamental challenge of balancing land use within a sustainable society.

The current economic system allocates value to land in a way that makes it very hard to maintain a stable and sustainable ecosystem. If we build on land, a rent can be charged, so it has value. If it is used for agriculture to grow cash crops, it has value. Land allowed to remain in its natural state is worthless. While the

5 An example of a third-generation biofuel uses algae to convert sunlight into an oil substitute. It requires a containment vessel, sunshine and some specialist processing, but can be located in any sunny location, including deserts.

world had a huge bank of undeveloped land, this flaw in the economic system was not apparent. Now the economic system must be changed to support a sustainable balance between human habitation and nature.

One method is 'natural resource accounting'.[6] This puts a value on natural land into a country's accounts. It then becomes a measure that can be monitored, prompting politicians to arrest any decline that shows up in the figures. This would be a general indicator. Politicians would then need specific policy measures to prevent further decline and support making improvements to reverse the decline that has already taken place.

One of these specific measures is the urban eco-balance tax. This links taxes on urban land with payments for natural land in order to change the economic parameters. The land the ecosystem needs then has value and an economic incentive exists to encourage good stewardship. The urban eco-balance tax is a flat-rate tax on all urban sites, based on the land area occupied. The funds raised are then used to support a payment system for all natural land. National parks would be outside the system, assuming that they are subject to watertight legal protection that keeps them forever outside the economy. This tax would be separate from, and additional to, other property taxes, although for administrative convenience it could piggyback on the same collection system. (Land used for agricultural production would incur neither tax nor benefit under this scheme.)

Owners of natural land would find that they could draw a yield by keeping the land in its natural state. The size of the payment would dictate the strength of the incentive, and some sort of inspection or certification regime would be required. Amongst the beneficiaries, farmers might find that the income from letting their land revert to natural habitats was comparable to the income from farming, but with less effort. In operating the system, vigilance would be required to ensure that this did not go too far.

For valuable urban sites in the centre of cities or densely populated communities, the flat-rate tax would be a relatively small additional overhead, so it would have little or no impact on the taxpayers.

6 Natural resource accounting is defined by the European Environment Agency (2010) as, 'A system of monitoring based on methodically organized accounts, representing the size of economically valuable and limited reserves of natural resources.'

For low-density urban sites, the flat-rate charge would be a significant overhead, so the biggest effect would be on urban sprawl. Urban eco-balance tax would compound the other negative factors of living in these areas, such as the escalating cost of commuting by car. As the attraction of living in these areas diminished, property values could crash. This would be the price of unwinding the twentieth-century model of urban expansion, but the regeneration that would follow would greatly improve the quality of community life.

I foresee developers spotting the opportunities and acting to regenerate such areas. The relatively large detached houses with wide drives and large gardens, particularly those of mediocre quality and poor energy efficiency, will be replaced. In their place, high-density, high-quality houses will be built around tight communities with excellent shared spaces. There are many reasons why such developments will dominate twenty-first-century construction (as discussed in Chapter 14): the urban eco-balance tax will play a small part in forcing the pace of change.

A further refinement to the proposed system could be to allow the income from any natural land to be offset against urban eco-balance tax charges, so that a balanced portfolio of ownership would become a part of sensible tax planning. Another effect would be on the layout of our urban spaces. For example, a developer rebuilding one of the old suburbs could reinstate a balance of natural land within the plan. An appropriate ownership structure could be used to balance the urban eco-balance tax with the payments claimed on the natural land, so each new resident would pay little tax as well as having access to a shared area of natural land where children could play with neighbours.

When I first published a proposal for a scheme to support the retention of natural land (McManners 2008), I went further to discuss international action to provide financial incentives between countries so that stewardship of natural habitats was rewarded. I believe that such action is vital and discussion should continue, but national action along the lines discussed here, implemented first in the developed countries, is the way to take the lead. This will set the principle that land left for nature has value and should be retained, not just in national parks but throughout the areas of land occupied by human society. This will be a double benefit in improving quality of life and the resilience of the ecosystem.

This chapter has been a brief review of green taxation and market mechanisms. Each of my three specific examples – the carbon market, localized supply chains and balanced land use – illustrates the concept, but there is more complexity in implementation than is discussed here. Taxation and markets are not appropriate in every case. Regulation may achieve a simpler solution. Markets are most powerful when there are resource constraints or pollution that needs to be controlled, and it is not clear how society should change to live within these constraints. Markets have the flexibility to harness human ingenuity in order to deliver the desired outcome. Above all else, it must be remembered that markets are subservient to the needs of humanity. A market has no purpose in its own right. It only has purpose if it delivers an improvement in world society or reinforces the integrity of the ecosystem.

PART 2

Changing Society and the Economy

6

Return to a Natural World Order

Making decisions on the basis of the sustainable policy framework will diffuse the worst excesses of economic globalization without turning away from the opportunities for further progress and improved human welfare. This is a natural order for world society in which humans and nature share a healthy coexistence.

Self-Determination

The effectiveness of the policy package lies in the power and balance that is intrinsic to self-determination. This is the way to mobilize people's selfish determination to protect their society and the environment. The process of change leads through three stages as sustainability is ingrained in society from the bottom up.

First, it is clearly in the self-interest of a community to protect the structure of its society and the environment in which its members live and from which they draw the resources they need. This is an easy concept to sell. Each country and each community takes responsibility for its own population, its own society and its own responsibility to coexist with nature.

From the basis of local sustainability, a society can move into the second stage and seek to broaden acceptance of the concept of sustainability to include accepting responsibility for the global impact of our actions. Communities with a strong grasp of their own sustainability have the intellectual maturity and knowledge to understand the impact that their decisions may have on areas beyond their direct influence. It is their choice whether to act on this insight by altering their actions. At first, there may be resistance to accepting constraints where the benefits accrue outside the community. Over time, we can expect sustainable thinking to become instinctive and move beyond the zero-sum game of shunting environmental stress to someone else's patch. The evident shared benefits will encourage cooperation and coordinated sustainable thinking.

In the third stage, enlightened self-interest leads countries to exert pressure to force other countries to adopt sustainable decision making. The result is a network of countries committed to sustainability. This should not be confused with localization. Sustainable communities may have a number of resource flows with other communities and regions, but this is with full knowledge of the totality of the environmental impact.

It seems implausible that today's society could be transformed to such an extent, but I believe that once such thought processes become ingrained, a sustainable policy framework becomes feasible. It would be impossible to convert the globalized world we have now from the top down. The connections (and disconnections) are too numerous and the complexity too great. Change has to come from the bottom up.

There will be inevitably countries that do not choose to follow a sustainable path. This will be especially true in the early stages when few countries are involved and the benefits not widely acknowledged. The countries that remain outside the expanding framework of sustainable nations may find short-term advantage but they are likely to suffer in the medium term as they attempt to draw resources from elsewhere but find it increasingly difficult.

Self-determination will meet with different levels of success, depending on a number of factors including population density and resource availability. Deprivation, discomfort and hunger will still exist but the search for solutions to such problems will take place within a safe and secure global ecosystem.

In a resource-constrained world, with communities and countries taking pride in determining their own affairs, it is inevitable that there will be conflicts. A series of small disputes will act as a safety valve to diffuse tension and bring resource consumption back into balance. This will be very different to an open globalized world market in which countries that can afford to pay get what they need and the poor miss out, until finally a breaking point is reached when the market unravels to bring widespread disruption.

Beyond Globalization

Proximization is a stage beyond globalization in which the imperative of sustainability is used to override raw global capitalism. This is not turning back history. It is progressing to a safe future, building on the best of our past achievements and backing off from the mistakes. The theories of free markets, deregulation and laissez-faire policy have been tried and tested and found wanting.

The key building block of this new world order is empowering people to work out their own way of developing a sustainable life for themselves and their community. This is how we can break out of the downward spiral of overconsumption of resources that is threatening to engulf the world, described by Garrett Hardin (1968) as the 'tragedy of the commons'. Each community and each country will take the decisions required to protect their resources and their natural heritage. Those countries with spare resources will be able to exert their power and influence to persuade other countries to adopt the same principles. The principle of sustainability will be implemented in ways that suit the circumstances, culture, capability and capacity of each country.

According to the WWF, in 1961 almost all countries in the world had more than enough capacity to meet their own demands (Hails 2008). The situation has changed radically, with many countries able to meet their needs only by importing resources from other nations.

We can quantify the situation we face by considering the Earth's capacity alongside the demands we are placing on it. Scientists have calculated that the Earth has 13.6 billion global hectares (gha)[1] of usable ecological capacity (land and shallow coastal regions). Countries and peoples of the world make differing demands on this resource. The average American requires 9.4 gha to support their lifestyle. This ecological footprint is considerably more than the 0.6 gha required by the average Bangladeshi. A typical European lifestyle falls somewhere in the middle: 4.7 gha. Australians consume 7.8 gha, while the Chinese consume 2.1 gha (Global Footprint Network 2008). Humanity's total global consumption is currently about 17.5 billion gha. This is 30 per cent more than the ecological capacity of our planet.

The true situation is likely to be far worse. Calculating a full set of ecological footprints for all the countries of the world is a time-consuming and complex process involving a considerable lead time. The figures quoted above were

1 Global hectares (gha) are hectares with world-average biological productivity (1 hectare = 2.47 acres). Footprint calculations use yield factors to take into account national differences in biological productivity. Footprint and biocapacity results for nations are calculated annually by the Global Footprint Network. The continuing methodological development of these National Footprint Accounts is overseen by a formal review committee (www.footprintstandards.org/committees).

published in 2008 based on data from 2005. Comparison with the preceding set of figures, which were published in 2006 and based on data from 2003, is worrying. The footprints of the United States and Europe show reductions of about 1 per cent. Although this is not a significant reduction, it indicates that the West has at least contained further expansion. In developing countries, ecological footprints are growing. Between 2003 and 2005, China's ecological footprint increased by over 30 per cent and India's increased by over 10 per cent. Over the same period, humanity's overconsumption increased from 20 per cent to 30 per cent more than the planet's ecological capacity. We do not yet have the results from more recent data, but there is every reason to believe that these trends have continued. Whilst we await the results of the calculations, there is nothing we can do alter the figures from 2005 to 2009. The data being gathered relates to facts we cannot now change. We should expect some very bad figures to be reported over the coming years and act now in anticipation.

It would be bad science to predict a figure for the world's overconsumption in 2009 before the data has been gathered and processed. But policy makers are disingenuous if they use the lead time to play down the risks. It is a fact that the world is overconsuming its ecological resources by at least 30 per cent, and it is a reasonable assumption that the true figure for 2009 will be considerably higher.

The combination of a lack of action by the West to make significant reductions and rapid development elsewhere in the world is severely overloading the planet. There has been much discussion over who is at fault, but the plain fact is that humans are overconsuming the planet and the pace of the consumption is increasing. Attributing blame is distracting effort that could be better spent. The West must accept that its model of development, based on high levels of consumption, is the wrong model for the world. The West must also accept a moral responsibility to develop and implement a different model that allows humanity to live sustainably.

Success will depend on the biggest countries adopting the principles of sustainability. These are the countries with the most potential to impact the global environment and the countries with the most influence over global society. Measured by CO_2 emissions, for many years the United States had the most negative influence on the global environment, until overtaken by China in 2006 (Netherlands Environment Assessment Agency 2008). How China and the United States decide to act will be critical to progress. These two countries are very different and will have to adopt different measures. The United States is consuming 90 per cent more than its ecological capacity, but it has sufficient resources within its own borders to become sustainable by halving its per capita resources consumption. This would require reducing average consumption from 9.4 gha to the footprint of an average European (4.7 gha).[2] For Americans this will be a big change of lifestyle, but one which seems entirely feasible from the European perspective. For China, the task is more difficult. Its huge population is already consuming more than double its ecological capacity, even though each person only consumes 2.1 gha.[3] The disparity between the situations faced by these two big countries reinforces the importance of using proximization to find sustainable living solutions. Policy that will work for the United States will not work for China, and vice versa.

Small countries also have an important role, not only to become sustainable societies to secure their own future, but also in terms of influencing global change. A small country, particularly if it has good natural resources, can lead in showing that a sustainable society is possible and is an attractive lifestyle. Sweden and New Zealand are examples, and they are leading the world in demonstrating sound environmental stewardship. Small countries can club together to use their voting power in world forums such as the UN to push for change. Some will lead in showing how to live sustainably; others will lead through explaining the dire circumstances they face because of the failure of the world community to act. For example, the people of the Maldives may be homeless before the end of this century as their islands are swamped by the sea.

Poor and populous ecological debtor nations are particularly at risk. Bangladesh is an example. Its circumstances are particularly difficult and troubling. The people of Bangladesh have one of the smallest ecological footprints on the planet (0.6 gha). The country also has a huge river delta, much of which is less than one metre above sea level. As sea levels rise, Bangladesh could lose one-fifth of its land area or more by the end of this century. Bangladesh will suffer as a result of rising sea levels that have been brought on by climate change, which in turn has been mainly caused by excess carbon emissions, mainly from the developed nations. This is evidently hugely unfair. The West has a moral obligation to help Bangladesh over the additional problems brought

2 US ecological footprint 2.8 billion gha; US ecological capacity 1.5 billion gha; US population 298 million; footprint per person 9.4 gha; average footprint per person in Europe 4.7 gha. If the United States reduced per person consumption to European levels the US footprint would drop to 1.4 billion gha.

3 China has a total biocapacity of 0.9 gha per person and a consumption of 2.1 gha per person, based on Global Footprint Network 2005 data.

on by rising sea levels. As governments and aid agencies search for ways to help, the solution will come from the delivery of social objectives (as discussed further in Chapter 8). There will be no place for imposing the Western model of development, or measuring progress by coarse economic measures such as GDP.

Measuring Progress

The management of a sustainable world society needs clear objectives. The old measures of development based on pure economic measures, such as GDP, are obsolescent. As the principle of sustainability embeds into world society, finance will drop down from being the prime objective of policy to a facilitating function. It will take time to develop appropriate measures such as National Resource Accounting (Harris and Fraser 2002) to bring ecosystem integrity on to the balance sheet. We will also need appropriate measures for society that include health and happiness such as the Human Development Index (HDI) launched by the UNDP in 1990 (UNDP 1990). Richard Layard (2005) explains the paradox that as Western societies have got richer, their people have become no happier. Bhutan's measurement of Gross National Happiness (GNH) (see p. 48) may not seem inappropriate as sustainable policy making takes root.

It will take time to unlearn the lessons of our materialistic upbringing, but, over time, many more people will learn to put quality of life before materialistic measures such as income and consumption. As the cohort of converts expands, people will find that they are judged by their contribution to the community and society. In communities, people are respected for the extent to which they are seen to contribute to the common good. The idea that the rich should be admired for being rich was always rather odd, but in a sustainable society this will be out of place. Tight communities naturally adopt these different measures. Financial wealth becomes a means for personal security, of course, but also a way to move up the hierarchy by paying back into society. Flaunting wealth and conspicuous consumption will go out of fashion. Making a donation to public facilities, which might then be named after you, will be a more acceptable and effective display of material wealth.

We tend to forget that such behaviour has always been part of society. For example, the philanthropist Andrew Carnegie is still remembered in a plaque on the walls of our local library. Carnegie the industrialist has been long forgotten. There is some evidence of a resurgence in personal measures of success rather than the simple measure of wealth. For example, Bill Gates' lasting legacy will be the Gates Foundation, not the personal wealth he accrued in his business career.

In measuring success, there is no need to aspire to a commune in which everything is shared equally. Differential benefits are acceptable when the community recognizes that this is the reward for delivering collective outcomes for the community. Senior leaders, managers and entrepreneurs can all earn materially more, without generating envy or resentment, if their efforts benefit the community, and this is seen to be so.

A New Sustainable World Order

A sustainable world will be full of variety: some societies may be dull and dependable, others vibrant and chaotic. There will be islands of paradise and sinks of despair. This is how the human world has always been. The struggle to live will continue, mixing pleasure with pain, and periods of hardship with periods of plenty. The new ingredient will be confidence that the generations to follow can enjoy the same struggle for life in circumstances of long-term macro stability.

The actions we take should hold society to the concept of sustainable living without stifling innovation. In the future, our inquisitive nature and drive for progress could take us to other planets and other solar systems, but this will not happen if those same human traits have destroyed the ecosystem and, with it, civilization. There is a credible risk that the complex web of society unravels in a struggle for resources in a globalized, overpopulated and overconsuming world (McManners 2009). This Armageddon alternative must be avoided.

In order to construct the new sustainable world order, society and the economy will have to change in tandem. The following chapters outline the parameters within key policy areas. The two most difficult challenges are population dynamics (discussed in Chapter 8) and developing workable mechanisms to implement green financial markets (discussed in Chapter 9). First, I will address the greening of global commodity flows. This is fairly easy to define but it will be very hard to implement, as the actions required will be seen as a reversal of previously successful policy.

7

Global Commodity Flows

Commodity prices can be volatile, rising when the world economy is strong and falling back in times of recession, but there will be no hiding from the relentless upward trajectory that is about to commence as the world comes out of recession. The rapidly increasing demand from fast-growing developing economies will not abate soon. Food prices, in particular, are being pushed higher. This is influenced by closer links between energy markets and agricultural commodity markets, as competition has increased for arable land to grow crops for biofuel production.

The open global economy has set up the impending resources crisis. Over the last few decades, an open global economy has kept many commodity prices low hence delivering improved economic performance. However, this has also created the illusion of plenty and prompted a rapid increase in rates of exploitation of non-renewable resources. From a conventional economic perspective, the problem is hidden; from a sustainable policy perspective, this is a glaring policy mistake.

The United Nations Conference on Trade and Development (UNCTAD) reported in their 2008 report (UNCTAD 2008) that 'the level and stability of commodity prices has become an important policy issue, not only from the traditional development perspective, but also from the perspective of the functioning of a highly integrated global economy'. The struggle for continued supply of commodities in an integrated global economy threatens to cause instability and high prices, and significant hardship, particularly for poorer countries.

The commodity markets have become highly sophisticated and are controlled through international exchanges such as the long-established Chicago Board of Trade and the London Metal Exchange. The New York Mercantile Exchange is the world's leading market for crude oil futures contracts, whilst the New York Board of Trade is one of the world's main markets for cocoa, coffee, cotton, wood pulp, sugar and frozen orange juice. Each commodity has a spot price and a range of future prices from near term to long term. In this complex market, the true purpose can be obscured behind speculation and arbitrage. The trade in commodities needs to return to the basic requirement of managing the demands of society in relation to the world's resources – and to do so in a sustainable manner.

The core purpose of trading commodities is to balance demand with resources. If a country needs more resources than are available within its own borders, it will seek supplies from another country that is willing to sell. Countries with spare capacity earn income from such sales. The value of the transaction is determined by supply and demand, tempered by the cost of transportation.

Open and unrestricted international commodity markets push prices down whilst supplies are plentiful. When demand exceeds supply, prices rise. The economic assumption is that investment in increased capacity will track rising prices. There may be a time lag before additional capacity comes on stream, but forward-looking extractive industries will attempt to invest early in anticipation of projected price rises. The futures market also guides farmers towards selecting the crop with the greatest profit potential. The counterparties in the market, such as corporations that need to consume commodities, reduce risk by hedging.

As prices rise, and investment flows, more mines are opened, more boreholes are sunk and more forest is cleared for agriculture. The market gives clear signals to support investment in machinery, facilities and prospecting for new deposits. This made economic sense in the relatively low-consumption world society of half a century ago. In the twenty-first-century world, which is already consuming resources at a greater rate than is sustainable, this economic logic is dangerous.

Overreliance on free trade will make it hard to maintain secure and sustainable supplies for society. For rich countries, such markets have worked well, so far. As demand outstrips supply, rich countries can expect to buy what they need but at a higher price. With the supply of commodities capped because we have reached the limits of what can be exploited, the poorest countries will not be able to get what they need. A poor country with limited resources will suffer greatly. A poor country that does have natural resources will seek to retain them for the use of its own population, or sell them only to another country that can supply key commodities in return.

Transforming the Commodity Market

The commodity market needs to be transformed so that protection of the environment and security of supply are given greater importance. As governments start to set policy in this way, restrictions on commodity flows will increase and, in many cases, prices can be expected to rise. Such consequences are the cost of delivering a secure and stable market.

In order to ensure a stable and sustainable international market in commodities, the policy that has been pushed hard for the last three decades needs to be reversed. Close oversight is required within a sustainable policy framework (see p. 152) in place of open markets and free trade. The default option is that resources are carefully monitored and conserved by the communities living close by, with priority being given to the needs of the local community before allowing exports.

In a sustainable market, tariffs are a useful mechanism. One of their purposes is to protect the country from imports to ensure the viability and continuance of local supply. Another purpose is to ensure resources are not overexploited, both locally and in regions beyond a government's direct control.

As the majority of rich countries adopt sustainable policy, a two-tier market may evolve: a main sustainable market and an unsustainable alternative market. Rich countries will use their influence to close down the alternative market. Countries that choose to be outside the 'sustainable club' may enjoy lower prices at first. Over time, most producing countries will transform their processes to gain access to the premium market. Countries that continue to purchase from the alternative market will suffer from the sanctions imposed by the countries committed to sustainability. The premium cost of buying commodities from sustainable sources will be the cost of entry to the world's premium markets.

Supporters of free trade will point out that, in general, prices on the world markets will be higher. This can be regarded as an insurance premium to ensure a safe environment and security of supply.

For some countries, prices in the internal market will be lower when they are living within their resources and a stable national market is maintained by releasing only excess capacity to the world market. These are likely to be resource-rich countries, many of which are underdeveloped and poor.

The countries that suffer from high world commodity prices will be those that consume well above their local resource base such as Switzerland and Japan. The governments of such countries will be forced to give priority to policy to reduce consumption – clearly a sustainable outcome. Japan is already a world leader in efficient resource usage with close cooperation between the Japanese Ministry of the Environment and the Ministry of Economy. A study by the Wuppertal Institute *Resource Efficiency: Japan and Europe at the Forefront* illustrates practical policy to promote reduced resource inputs (Bahn-Walkowiak et al. 2008).

Instead of a global market that uses price signals to drive investment in activities that deplete natural resources, the market would drive investment in activities that minimize resource usage. Such a market works in favour of sustainable policy. Policy makers have to look outside the narrow conventional economic perspective, and beyond the arguments in favour of free trade, to understand these effects.

Sustainability as the Bedrock of Policy

The default option for any sustainable community or country is to live within its resource base. The natural consequence is fewer imports and, usually, fewer exports, depending on the circumstances of the country. In considering how best to utilize local resources, thought must be given to whether it is possible or sensible to implement a local sustainable solution. A blinkered approach that local is always best is not necessarily the correct course of action.

Security of supply

A policy that relies on imports, when continuation of supply is uncertain, is risky. A policy that ensures security of supply must look closely at the source of commodities. Are the source countries reliable partners? Are they likely to continue to have excess capacity? For example, there are big exporters of commodities which have considerable power in the market. It should be considered whether such countries are likely to pull capacity off the world market in order to satisfy their domestic demand or gain political leverage. In a resource-constrained world, reliance on imports should be backed up by robust bilateral or multilateral agreements to guarantee supply.

In considering security of supply, chronic overconsumption may be identified, in which case policy should be carefully crafted to address the cause and reduce consumption to sustainable levels. It is far better to do this early, on a country-by-country or region-by-region basis, than waiting until global shortages lead to escalating prices and a scramble for supplies. Waiting until price signals show an urgent need to act will make the transition to lower consumption much more difficult than it needs to be.

Reducing the environmental impact

The environmental impact of the extraction and production of commodities is a problem that is often exacerbated by international trade. The WTO Committee on Trade and Environment (CTE) was established in response to this problem, but its founding agreement (WTO 1994) states that 'there should not be, nor need be, any policy contradiction between upholding and safeguarding an open, non-discriminatory and equitable multilateral trading system on the one hand, and acting for the protection of the environment, and the promotion of sustainable development on the other'. In effect, this rules out examining whether the policy of free trade and sustainability are compatible. It is not surprising that the CTE has not recommended any changes to the rules of the multilateral trading system (WTO 2009b) – although the CTE has raised issues that have migrated into the negotiations of the Doha round of world trade talks, such as the relationship between the WTO rules and multilateral environmental agreements (MEAs).

Trade policy needs to bring environmental safeguards to the fore, so that potential problems are anticipated and prevented, rather than waiting for symptoms to show before discussing action. This will require open and well-informed dialogue in which all options are considered. It is not helpful to place the CTE in a straitjacket that defends the status quo.

The internal perspective

Countries have good reason not allow external demand for resources to adversely affect the local environment. It is evidently in a country's self-interest to ensure that extractive industries, agriculture and forestry are managed to protect and conserve the local environment. Where policy to protect the environment is failing, due to, for example, corruption or weak enforcement, governments should consider imposing tight controls on exports. Even a government with weak control over its internal affairs is likely to be in control of its borders, so can take action through this route despite the domestic difficulties. Tight control of exports would remove the prime incentive to circumvent national environmental policy.

Another issue that may arise is that local resources are intrinsically unsustainable. The sustainable solutions are either to operate society without consuming the resource (the most sustainable solution), or to look elsewhere for alternative supplies. For example, if low-grade coal is the only fossil fuel available locally, policy should be pushed very hard to move to a zero-carbon economy. During the interim period, it would make sense to import gas – a much cleaner fuel. A country with its own gas supplies may not act with such urgency to build a zero-carbon economy, knowing that, as gas reserves run low, it can stop exports to retain the gas for its own consumption.

The external perspective

It is now understood that environmental protection of the local area offers no escape from changes to the global ecosystem, such as climate change and rising sea levels. A sustainable policy framework must, therefore, include the principle of acting to influence sustainable behaviour outside national borders.

The source of imports must be reliably documented and an assessment made as to whether it is sustainable. For example, timber should come from sustainably managed forests; agricultural products should come from countries committed to preventing further destruction of natural wilderness (such as rainforest); and metals should come from countries that employ environmentally safe extraction techniques.

The external leg of a sustainable trade policy could be disadvantageous to the countries that are doing the most to influence the adoption of high environmental standards. Importing only from sustainable sources is likely to be more expensive than buying at lowest price without worrying about the consequences. This will certainly be true in the early stages, before sustainable policies are implemented around the world. Later, when sustainability is understood and countries that disregard the environment find that they are shut out of most markets, the situation will settle into balance.

During the transition, the countries that lead in setting sustainable policy will also lead in developing more efficient, cleaner procedures and processes. This network of countries can be expected to influence the market and make things difficult for countries that remain outside the network.

There is considerable potential for disputes in the sustainable trade of commodities. There would be a useful role for an international body to discuss measures and broker solutions. Adding measures to supplement the existing WTO rules will not be sufficient. Policy needs to shift from the defence of free trade to the promotion of sustainable trade (as discussed further in Chapter 11).

The concept of sustainable trade can be illustrated by a discussion of the market for a number of key commodities.

Food

Food is the most basic commodity of all. The default sustainable option is to grow our food close to where we live. This provides the circumstances for fresh, healthy food. Local agriculture is subject to oversight by the community, which is far more effective than any system of official inspection. Reliance on local agriculture also gives communities some control over security of supply and a gut understanding of the risk when the community consumes more than the available agricultural capacity.

Poorer countries will be more exposed than the rich developed countries. When demand outstrips supply, prices will rise and the poorest nations will struggle to afford the food they need. Restrictions and tariffs to protect the local market are sensible in order to insulate it from fluctuating world prices. Protection is necessary when world food prices are low and cheap imports threaten to destroy local capacity. (This is the justification used by many countries to provide financial support to agricultural producers, which in the case of Switzerland, Iceland, Norway, Korea and Japan is over 60 per cent of gross farm receipts (OECD 2009).) Protection is also needed when prices are high and there is a risk of staple foods being sucked out of the economy by more affluent countries.

The world demand for food is set to rise owing to a growing number of mouths to feed and increasing affluence leading to a desire for richer diets.

World agricultural capacity will struggle to cope, especially if robust policies to protect the rainforest and other wilderness areas are successfully adopted. These are areas that could be brought into agricultural production but, according to the tenets of sustainability, it is vital that such a response is resisted.

World food production will not, therefore, match worldwide demand. In an open globalized market, poor countries go hungry and are dependent on charitable donations. Localized markets will manage this imbalance much better. Countries that are not self-sufficient will try to become less reliant on imports. Improved agricultural efficiency will be one element in restoring balance. Another will be that foods requiring a large ecological footprint for their production will be much more expensive. Market forces will encourage reversion to diets with less environmental impact, such as vegetarian or low-meat diets.

The challenges of ensuring a sustainable food supply will oblige policy makers to confront the issue of population growth. Whilst there is the belief that an open world market can always provide food supplies, we do not perceive this to be a pressing problem. As policy makers understand the growing difficulties of ensuring secure, sustainable food supplies, the challenge of formulating policies to constrain the size of the population may then seem more urgent.

The transition to local agriculture, as the preferred option, will apply to all countries, both poor and rich. Many people would be content to confine eating strawberries, for example, to the summer months. People who insist on eating fresh strawberries out of season should pay a high premium. The poor and less well off will have their choices steered towards seasonal local produce. The distance-to-market taxes outlined in Chapter 5 will lean most heavily on exotic fresh foods consumed by those who can afford to pay, not on the poor and needy. Politicians would be justified in setting such taxes higher than the cost of the environmental impact. If the principle of assigning taxes to related expenditure is applied, such taxation could be allocated to pay for a universal benefit, such as free school meals.

Sustainable food supplies will also be healthier, because they will be fresher, safer and will steer us away from overconsumption of meat. It will also be far more difficult for industrial processes to undermine the market and affect our health.[1] Sustainable food chains are substantially different to current

1 For an account of the risks and dangers of the modern industrial food supply chain, see McManners (2009: 165–82).

commercial practice. There are many examples. I single out milk consumption in the UK, not because it is a particularly serious problem, but because it shows a fundamental lack of understanding of what sustainable policy should entail. In a recent report to the UK Government, aimed at examining the environmental impact of the milk supply chain, the option of shortening supply chains was ruled out. It is astonishing that the commitment to economic globalization within modern society is so great that such an obvious solution was not even considered (see Box 7.1).

BOX 7.1 THE MILK ROAD MAP 2008

In 2008, the UK Government published *The Milk Road Map* (DEFRA 2008). This was produced by The Dairy Supply Chain Forum's Sustainable Consumption and Production Task Force, a group of dairy-industry representatives and government agencies including the Department for Environment, Food and Rural Affairs (DEFRA).

The aim of the report was to examine the environmental impact of milk production, supply and consumption. The task force based their analysis on a key underlying assumption which they, and the government, accepted without question:

'The globalisation of dairy markets implies that there will be further rationalisation of dairy processing businesses, in order to both achieve greater economies of scale and secure market presence in the key global markets with increasingly global retailers.'

The milk road map goes on to identify a whole range of possible measures to reduce the environmental impact of the milk supply chain, ranging from using more fuel-efficient vehicles to improving the recycling of milk containers. It is an excellent report in many ways, but demonstrates a mindset that has become deeply rooted. The assumption that there will be further centralization of milk processing and further expansion of the global trade in milk and milk products acts as blinkers that prevent policy makers from seeing the most obvious sustainable solution. There is no need for such long supply chains, which include large consignments of yoghurt crossing Europe in opposite directions under different brand names. The simplest and most sustainable action would be to adopt policy that favours dismantling long supply chains and encourages local production and consumption.

Food is not just another commodity: great care must be taken to ensure that it is safe and that supplies are sustainable. Policy makers have to learn and understand a different approach. An argument can be made for the retention of an open world market in long-life staple foods, such as maize and grain, which can be transported easily and stored for long periods. This argument is based on the need to make food available to countries that have suffered a poor harvest and need supplies to feed a hungry population. In isolated instances, such as a drought in one country, an open world market can help, but it is dangerous always to rely on supplies from an amorphous global market on the assumption that rising prices will pull capacity from somewhere. Such a globalized market runs the risk of an occasional severe global food crisis. The key to sustainable food supply is careful husbandry of reserves at the local or regional level, so that localities remain in control of their food security and can plan accordingly.

Energy and Fuel

Concern over climate change and the need to reduce CO_2 emissions has generated considerable discussion over what is to be done. The current emphasis is on brokering global agreements to limit emissions and using carbon markets for implementation. This may achieve reductions in the short term as world consumption shifts in favour of the relatively clean fuels such as gas. As these reserves run low, a stable carbon market will help in planning investment to exploit poorer quality reserves such as coal and oil sands. These efforts may delay necessary reform and lead to greater CO_2 emissions over the long term. (See Chapter 5, pp. 64–5.)

The situation requires much bolder action. The energy market has to be flipped. Instead of regarding demand for energy as a need to be met, leading to building ever more power stations, we have to turn to renewable energy. This requires a reversal of policy. Each locality has a feasible renewable energy supply and communities have to learn to live within this constraint, supplemented only by sources that are proven to be sustainable.

In the short term, the world is addicted to fossil fuel. The transition away from fossil fuel is a complex challenge, but it is entirely feasible (McManners 2008: 51–68). It is worth focusing on the future of sustainable energy systems in order to be able to embark quickly on the right path.

The world's immediate response has been policy to favour biofuel.[2] We now know that first-generation biofuels[3] are not the solution, because they divert

2 The US renewable fuel standard aims to achieve 7.5 billion gallons of biofuel blended into gasoline by 2012. In January 2008, the European Commission proposed a 10 per cent binding minimum target for biofuels in transport by 2020.

3 First-generation biofuels are extracted from crops, such as rapeseed or palm oil to produce biodiesel, and grain or sugar cane to produce ethanol.

agricultural capacity from food to fuel. In a hungry world, this is not useful. Second-generation biofuels from organic waste are much more useful. These biofuels, together with wind, hydro, solar, tidal and geothermal power, are the foundations of our future energy budget (until developments in nuclear fusion in the long future make energy cheap and abundant once more).[4]

The international trade in fuel and energy will change. Energy generation and consumption needs to become much more localized. My research also indicates that, to make real reductions over the medium and long term, we will have to find a way to ban the trade of coal and low-grade fossil fuel. This is highly controversial and I accept it will be difficult. There would be some localities where, in the short term, there will be no choice but to use such dirty fuels, but more and more countries must be persuaded to close down low-grade fossil-fuel extraction. This country-by-country approach may work if action is taken before reserves of cleaner fuel run low. If the fossil-fuel market remains an open global market, I see little prospect of reducing consumption within the timescale recommended by scientists (IPCC 2007) to prevent serious consequences for the planet and humanity.

One challenge is that the renewable power available is dependent on location. Countries at high latitudes with little solar power and long cold winters will struggle to cope (unless they are very sparsely populated and have large forest assets to draw upon). Shutting down the trade in fossil fuel will cause real hardship. The answer lies in an international market in 'sunshine' exported from regions where there is abundance. This may be transported as electricity or 'liquid sunshine'.

I use the generic term 'liquid sunshine' to avoid second-guessing the best technological solution. It is likely that liquid sunshine will be a biofuel produced by biotechnology or other processes, yet to be invented, that mimic photosynthesis. Another possibility is that liquid sunshine will be liquid hydrogen. It depends on which process delivers the most cost-effective product. In 2010, oil is so cheap that none of the options are viable. As energy prices rise, we will reach a point where there is a robust business case for liquid sunshine.

4 Viable nuclear fusion plants are estimated to be 30 or 40 years into the future. Conventional nuclear power is not mentioned here. The debate for and against energy from nuclear fission is finely balanced. Those people in favour have a narrow focus on making CO_2 emission reductions; those against believe the long-term legacy and risks are too high.

The world has large areas of desert where solar radiation is sufficient to give a huge boost to the world's renewable energy budget. These areas are not used for agriculture so there is no food-fuel trade-off. All that is needed is the technology and the investment. The technology of third-generation biofuels, produced by algae grown in enclosed tanks, is promising. Solar power stations also show promise,[5] particularly in locations close to population centres, such as Spain and North Africa where energy could be generated for European consumers.

Power-hungry processes, such as smelting aluminium, will migrate to places of abundant renewable power. This transition has already begun. Iceland, with its huge hydro potential, is already becoming a major world centre for aluminium smelting, but it has to compete for business with other aluminium smelters, such as those in China which use cheap power from coal-burning power stations. New sustainable trade regulations would penalize aluminium made using energy from coal compared with aluminium from Iceland.

Trade in sustainable energy will be dominated by local markets for renewable energy supplemented by international trade in 'liquid sunshine'. There will also be international trade in commodities with a high energy input which have been produced in places with ample renewable energy

Consumption and Production

Sustainable production is based on closed material cycles described by McDonough and Braungart (2002) as cradle-to-cradle manufacturing. This involves a different set of processes to those used in current manufacturing but it is entirely feasible (McManners 2008: 117–24). Implementation will lead to changes to world trade.

The key requirement is a close relationship between manufacturer and consumer. Careful examination of sustainable production leads to the migration of processes from simple one-off sales to an ongoing relationship with the manufacturer. For example, the world market in domestic white goods is unsustainable. Products such as washing machines are bought and used for between 10 and 20 years before being discarded. The sustainable alternative is

5 Solar power stations rely on two main technologies. One is an array of mirrors focusing sunlight on to a liquid containment vessel to generate steam to drive turbines. Another method is vast arrays of photovoltaic panels. These technologies vie for the least expensive solution.

a sale or lease with the machine being repaired and refurbished throughout its lifetime by the manufacturer, who then takes it back to be fully recycled.

The global supply chains of today are designed to maximize the efficiency of the simple sales model, and act as a barrier to establishing sustainable production and consumption. The transformation required is to shift to small-scale production facilities located with easy access to consumers, including full product support and end-of-life refurbishment or recycling.

Relatively small flows of international trade in physical goods will remain when these are specialist high-value items, such as large aircraft, computer chips, advanced machine tools and robots. This is due to the complexity of manufacture as well as sensitivity at sharing advanced technology within the global knowledge economy (see Chapter 10) and the need to protect intellectual property rights (IPR).

The sustainable future of manufacturing entails localized production within closed material cycles that require much reduced inputs of raw commodities. The regulations to support the transformation will lead to a major contraction in the world trade in physical goods.

Waste

As regulations for waste management are tightened, waste destined for disposal becomes a valuable commodity. Taking responsibility for waste will become increasingly profitable.

During the economic boom of recent decades, concerns about waste-handling have gone no further than compliance with national regulations. There has been no need to think any more deeply than shunting off waste elsewhere, to a waste contractor or municipal rubbish system. One way to circumvent onerous national regulations is to transfer dirty industry to other countries. This need not be deliberate policy – it is a consequence of the world economic system based on free trade, in which dirty activities migrate to where regulation is least strict. The resulting pollution often ends up in the shared atmosphere or ocean. The resources of our planet, and its capacity to digest waste, are limited. We must treat the planet as if it were our own back garden. We need to bring 'elsewhere' with regard to waste much closer to home.

The most sustainable (and often most profitable) solution is to change processes in order to eliminate waste. When waste is a valuable commodity, it makes economic sense to invest in research and development to design waste out of industrial processes. When this is not possible on a process-by-process basis, an alternative approach is to link whole networks of activities, with the output from one process becoming an input to another. This can apply right across society, from industrial processes to household consumption. The throwaway society can be transformed into a society where true recycling is universal and waste management systems, including domestic refuse collection, become obsolete (McManners 2008: 149–56).

The international trade in waste is particularly dangerous. When waste leaves our borders, we lose control. Regulations require a certificate of safe disposal, but this may be available for a fee that has little connection with safe disposal. It is easy to 'lose' the cargo in transit on the high seas, provided that the dodgy paperwork is good enough to survive scrutiny. Alternatively, the waste can end up being dumped in a remote corner of the world where people have little idea what has arrived and even less knowledge of how to deal with it safely. Of course, responsible governments and corporations will not behave like this and insist on safe disposal. It is my opinion that the world will agree, in the end, that the only fully sustainable solution is to ban the international trade in waste.

Conclusion

A shift from free trade to sustainable trade is needed to deliver sustainable solutions to providing the resources that society needs. This is a major reversal of policy and it requires a shift of mindset away from the current ruling logic that trade between countries 'is almost always to their mutual benefit' (Krugman and Obstfeld 2009). The green economist Volker Heinemann (2007) puts forward a detailed argument that continued adherence to free trade is 'out of touch with reality'. He concludes that the existing international trade bodies should allow for specific economic policies that manage international trade. However, he also notes that 'nobody knows what really works in international trade and how the underdevelopment in this world with related overpopulation and environmental problems can be solved'.

The key conclusion is that free trade is no longer the appropriate basis on which to manage international trade in commodities (and goods). Finding alternative policies requires sound economic thinking to implement sustainable trade policy. Economists have a lot of work ahead being careful to avoid being blinkered by conventional concepts.

My contribution to the debate on the commodity trade flows is based on an analysis across a wide range of issues, some of which have been covered here. I encourage any reader who remains sceptical to read the references in this chapter to follow the logic behind the argument. The negative aspect of the policy, where resistance will focus, is that commodities will cost more over the short term. This will particularly affect the rich nations most reliant on imported commodities. It will be a difficult process of persuasion to explain that this is a price worth paying for long-term security of supply and to conserve resources for the future.

Stripping out the planet's natural resources is a loss for everyone; releasing pollution into the environment affects us all. The concept of 'elsewhere' with regard to both resources and pollution is a dangerous excuse to disregard nature and do anything we please. The alternative approach is for every community and every country to take care of the activities in their own backyard by stopping unsustainable exploitation and placing tight controls on polluting activities. This would provide the circumstances for a sustainable world. From within the security of sustainable societies, we can then influence other countries to tend their patch of the shared ecosystem with the same diligence. This is sensible, feasible and, once understood, acceptable to the majority of people. The more difficult challenge is to bring the same logical thought process to bear upon population dynamics.

8

Population Dynamics

The world has embraced free trade and free flows of capital but not the free movement of people, as a perfect free market would require. There are reasons for this. Governments are responsible for social provision and cannot afford open borders. Barriers to immigration are a natural consequence.

The proposals in this book will lead to less migration of people, not because of draconian regulation, but because the world works better when people are engaged with their locality and empowered to build a better life for their family and the community around them.

The altered population dynamics that accompany a sustainable world have the benefit of providing the circumstances that will allow population growth to be brought under control. This is not the prime aim of setting sustainable policy, so it may seem to be a fortuitous side effect. If we pause to think a little more deeply, however, it becomes clear that this is not just a side effect but a logical consequential outcome. It is impossible to have a sustainable human society on this planet unless population growth is brought under control.

For world politicians, the issue of population, and the need to control it, is political dynamite; for authors, tackling the subject risks causing offence. We need a debate amongst compassionate and sensible people because this issue is the 'elephant in the room'.[1]

Population Growth

Population growth is an emotive subject that reaches deep into our inner psychology. When writing about it, it is usual to choose words carefully. This

1 I attended a lecture by a leading veteran politician about sustainability and government. Under the protection of Chatham-House rules, he explained that population is the 'elephant in the room' that politicians dare not address because it is 'political dynamite'.

allows us to navigate around the issues without causing offence. Such niceties of language have a cost attached. The issues become obscured to such an extent that we can miss the main point, which is contained in the equation:

Population × Individual Consumption = Load on the Earth's systems

The current population is consuming at a rate that is 30 per cent more than the planet can sustain. Each year that we allow this excessive load to continue, the Earth is suffering. It will take time before people in the developed world experience the results first-hand. As the demand for agricultural products and other commodities increases, prices rise, and we complain, but we can afford it. Suffering is confined to the poorer countries and the less well-off elements of society. This is not an equitable and fair way to run world affairs. As we continue to overspend on the resources account, we will strip the Earth bare. There is no denying the truth of the mathematics. We must bring this equation back into balance.

The two terms on the left of the equation have to become smaller numbers. Population and/or consumption have to reduce.

Focusing on consumption is the less sensitive issue. There is a multitude of ways we can change our lifestyles to consume less of the Earth's resources. For example, living in small compact communities is a much less resource-intensive model of living. We could also eat less meat, use public transport to replace private car use, live in well-insulated houses, manufacture fully recyclable products and the list goes on.

Making substantive reductions in consumption in the developed world will cut deep into the lifestyles to which we have become accustomed. Many of these are necessary adjustments but on their own will not be enough.

Looking at the equation as a whole, we notice that a narrow focus on consumption is not going to help us much. The world population is continuing to rise, so consumption has to be driven yet lower to compensate. No matter how low we can drive per-capita consumption, a continually growing population will exceed the Earth's capacity. That is the unbiased logic of the maths. There is no avoiding the issue of population growth.

The twentieth century was a period of extraordinary population growth from 1.6 billion to 6 billion. United Nations' estimates indicate that world population will reach 7 billion by 2012 and 9 billion by 2050 (UN 2009). This continued expansion is not sustainable, especially if per person consumption is also increasing. Population growth and conventional economic development is a dangerous combination as countries compete for the planet's diminishing resources. The imperative to bring population growth under control is growing ever stronger.

The debate over population growth requires great care. The discussion soon enters into issues of personal choice, freedom, human rights and equity. This chapter does not attempt to solve all the problems associated with population growth. Rather, it highlights the contribution that policy for a proximized world can make to improving population dynamics.

Economic globalization has undermined the natural mechanisms that tend to keep populations in check. When resources are capped there is an instinctive human behaviour holding population growth back. The illusion of limitless supply of an open global market has dulled this subtle mechanism. An equally subtle reverse effect can apply to expand population in places where family size increases economic wealth.

Proximization reintroduces the linkage between people and their resource base and plays a part in rebuilding the delicate dynamic balance of human populations.

The Movement of Labour

The economic argument in favour of the free movement of labour is clear. Labour migrates to where there is work. Market forces match up people looking for work with vacancies in the most efficient way. Companies get the skills they need at the lowest cost and the individual is free to sell their expertise to the highest bidder.

In the rich developed countries, opening up the labour market puts considerable downward pressure on wages, particularly for unskilled or semi-skilled jobs that require little education. The power of unskilled workers to hold out for higher pay is much reduced when the labour market is open to a pool of international labour. This brake on labour costs allows the economy to run more efficiently. The associated immigration policy is, of course, far more complex than this simple economic model. Politicians have to balance the economic benefits of immigration with the increased load that it puts on the country's infrastructure and resources.

In poor developing countries, the effect of open international labour markets is different. There, the attraction of high pay in developed countries is a magnet to the whole population but, as developed countries restrict immigration within manageable quotas, it is only the most able and best qualified who get through the selection process. Poor countries can find that they lose doctors, nurses, engineers and other professionals that the country badly needs.

Policy decisions on the international movement of labour must balance the short-term economic benefits with longer-term economic outcomes, as well as social provision and the sustainability of society.

Automation and Labour

Automation replaces menial work and allows each person to have far more possessions and consume far more than society could otherwise afford. This progress comes at a cost. The increasingly automated modern world needs fewer workers to achieve the same output. This applies to, for example, factories with increased use of robots, retail operations with automatic checkout facilities, service industries with web-based self-service systems, and even professionals whose advice can be replaced by computers with artificial intelligence. The point will soon be reached when almost anything can be automated – if we so choose. As the machines take over, humans will be left with management roles and highly skilled jobs that are hard to automate. In the future, the choice of what should be automated will become very important and will be more than a simple decision over direct costs.

The conventional economic response to spare capacity in the labour market is to boost consumption. It is assumed that this will lead to more jobs. This assumption is starting to break down. In an increasingly automated world, boosting consumption will lead to more robots, and more resources consumed, but not necessarily more jobs. Even robots can be made by other robots in factories controlled largely by computers, with a few senior managers and a small team of maintenance staff to deal with unusual or unexpected situations. As each new problem is fixed by human ingenuity, the solution can be entered into a knowledge database so that the next time the same problem occurs, the robots can fix it.

This is not science fiction. The developed world is entering an era of advanced automation that will require a different attitude to the labour market.

Governments have a social responsibility to the population as a whole. A situation in which there is a growing imbalance between a small group of senior managers and experts and a large underclass of the unemployed is not sustainable. Even an affluent society that can afford to pay reasonable social security payments is undermined by high levels of unemployment that put social cohesion at risk.

Governments have a responsibility for society, and every person in that society. Policies will be required to ensure that a sustainable society, even one with access to the latest high technology, engages the population in work. One possibility is to accept that each person of working age is a potential liability on the state and use this as a factor in all decisions. This will affect the business case as to whether to automate or not. The decision to implement systems that eliminate jobs (particularly if these are within the public sector) has to factor in the increased cost to the social security budget. The converse case can be made through designing systems that require more people and save on social security expenditure (McManners 2008: 181–91). This alternative case should be considered alongside other options even though it conflicts with the widely accepted management theory that automation should be used to reduce head count. The key challenge is to learn this new way of thinking without sinking into Soviet-style inefficiency or creating jobs that have no purpose.

I have experience of writing business cases for IT systems as a practitioner and through teaching MBA students over the years. The underlying assumption of many simple business cases is to eliminate jobs, saving expenditure on wages. To think in reverse and consider whether capital expenditure can be saved by employing more people is not an easy concept for managers.

It may not be necessary to deal with the challenge of automation in the next decade or two. There are many aspects of creating a sustainable world that need considerably more effort and ingenuity. So when bringing forward legislation to require sustainable processes, such as full recycling and cradle-to-cradle production, policy makers should be aware of the beneficial effect of generating employment. In the short term, committing to transform society to become sustainable may be enough to counter the threat of excessive automation.

Resources and Population

Adopting sustainable policy exposes the conflict between the size of a population and the level of consumption. The size of a population, and the consumption per person, are not necessarily problems if taken in isolation. A country with a wealth of resources and a small population, such as New Zealand, could adopt sustainable policy with relative ease. A highly populous country, such as Bangladesh, will find it very hard, especially as it is already consuming more resources than it has within its own borders. The mathematical equation that defines the situation is:[2]

$$\text{Population} \times \text{Individual Consumption} = \text{Total Consumption}$$

For a country to live sustainably, the total resource consumption must not exceed the sustainable resources available. If there is a deficit, the maths tells us that either average individual consumption must reduce or population must reduce. This is a simple harsh reality. It is better to face the reality early than find ways to fudge the issue.

A country's ecological capacity, compared with consumption, provides a good test of whether long-term sustainability is feasible. A surplus or deficit is not a fundamental measure of sustainability, as sustainable flows of trade can be used to cover imbalances. However, in a world that is already overconsuming natural resources and demand is rising, governments will be taking a big risk in relying on trade to secure resources into the future. More than three-quarters of the world's people live in nations that are ecological debtors (Hails 2008). As this trend continues, borrowing ecological capacity from other nations will become harder.

Some examples of regions and countries in surplus and deficit are shown in Table 8.1.

The countries with the biggest ecological debts are China (-1.6 billion gha), United States (-1.3 billion gha), Japan (-0.6 billion gha) and India (-0.6 billion gha). The countries with the biggest ecological surpluses are Brazil (+0.9 billion gha) and Russia (+0.6 billion gha). Worldwide, the surpluses are not enough

2 It has been suggested by environmentalists such as Paul Ehrlich (1968) that an equation that can be used is: $EB = P \times A \times T$, where EB = environmental burden, P = population, A = affluence and T = technology (reported by Hart 2005). This is used to support the view that population and affluence can grow provided the technology to create wealth is changed fundamentally. I believe that this obscures the core issue so I prefer the clarity of the simpler equation.

Table 8.1 Ecological deficits and surpluses

Country/region	Population (millions)	Ecological footprint per person (gha)	Ecological surplus (+) or deficit (-) (millions gha)
World	6,476	2.7	-3,886
Europe (EU)	487	4.7	-1,170
China	1,323	2.1	-1,588
United States	298	9.4	-1,312
Japan	128	4.9	-551
India	1,103	0.9	-552
Brazil	186	2.4	+913
Russia	143	3.7	+630

Source: Global Footprint Network (2008).
(See p. 77 for the definition of gha.)

to offset overconsumption. Overall, the world is running a deficit of 3.8 billion gha. The figures show that the priority, from a world perspective, should be to find ways to persuade the biggest debtors to change policy.

For the United States, the task is relatively easy. If the United States were to reduce per person consumption to match the European average the country could be sustainable, living off the resources within its borders (as mentioned in Chapter 6). From the perspective of a resident of Europe, this does not seem too hard. In general, Europeans may not lead such consumerist lives as Americans, but Europeans live well. If the United States chose to become sustainable, the result would evidently be good for the planet, and should attract wide support. It is worth noting that this would also mean that the United States would no longer be exporting huge amounts of agricultural produce to other countries. Countries that have become reliant on such supplies need to be aware of this.

For China, the challenge is different. In the 1960s, China still had a low-impact agrarian economy and was living within its ecological capacity. In the early 1970s, China moved into deficit as it started to develop an industrial economy (Kitzes et al. 2008). Consumption in China is growing fast, with the main component of the deficit being increased consumption of fossil fuel. Per person consumption is low compared with the leading developed nations. If China continues to aspire to develop following the example of the developed nations, it is hard to see how China can bring its ecological deficit back into balance.

$$\text{Population} \times \text{Individual Consumption} = \text{Total Consumption}$$

Looking at the fundamental resource equation, there is no way to avoid the logic that China has a problem with population. This is an uncomfortable truth that is often avoided.

Bangladesh is another example. The country is not a major contributor to world overconsumption but it has a huge problem if it is to live sustainably in the future. The fact that it will lose land area due to rising sea levels is not a problem of its own making, and the world community has a moral responsibility to help. This should not be allowed to obscure the fact that Bangladesh is already consuming double the resources that are available within its own borders. Individual consumption levels are amongst the lowest in the world (0.6 gha). Only the people of Malawi and Afghanistan consume less, according to the Global Footprint Network (2008). It is not feasible to reduce such a low figure. The priority has to be policy to tackle population growth.

Shifting to sustainable policy highlights the issue of resources and population in a way that forces policy makers to address the issue. This is different to a world of open borders and free trade, which tends to gloss over the issue of excessive population. While there still is enough world capacity, the awkward issue can be avoided.

Reducing Population

The incentive to develop and implement sustainable policy frameworks is high because of the consequential effect on population numbers. Once a country has policies that can bring population under control, it becomes feasible to consider not just stabilizing population but also reducing population. This opens up the prospect of improving human welfare through development that increases consumption. It is possible to live within capped resources through a combination of reducing population and growing consumption. This will make it easier to build balanced sustainable societies.

It is sometimes argued that economic development that increases consumption should be the first stage in improving poor societies. This is based on the assumption that a more affluent society is better able to control population. The low birth rates in rich Western countries seem to bear this out. However, this is a risky strategy because if the assumption is wrong the situation can become far worse.

There are policy choices available that directly target population control. These can work even in the poorest countries and should precede or go in parallel with other development efforts. Examples of such policies are linking communities to their resource base, improving infant mortality, providing reliable care for the elderly (that is appropriate to the community), ensuring that having children incurs costs (at an affordable level) and improving the education of girls and young women (McManners 2009: 73–83).

There is no avoiding the sensitive issue of population. When policy makers examine how to make a society sustainable, the first factor to consider is the resources available locally and the opportunity to replace shortfalls with imports from sustainable sources. If the imbalance between demand and sustainable resources is large, reductions in consumption are required. If reducing individual consumption cannot deliver the required reduction, then policy must focus on reducing population.

Emigration

Sustainable policy will lead to restrictions on immigration as countries work on policy to secure resources for their existing population. Successful sustainable policy will also put a brake on emigration. People have a natural affinity for the place where they are born and raised. As people take more pride in home, culture and place, there will be less need to consider leaving.

There will still be envy at the apparent success of other countries, and there will be countries that are so hopelessly incompetent and corrupt that people have good reasons for wanting to leave. However, human identity is so tied up with family and tribe that the first choice for most people is to seek to make the best of their home. Sustainable communities are built on this human instinct to support the people close by. Instead of envy at the way of life in other countries, it would be better to encourage pride in the way things are done at home.

Cultural globalization, facilitated by global brands and TV programmes, has tended to work against building pride in a country's own identity. That is why maintaining the cultural identity of communities is so important. This is not just to do with preserving heritage for posterity – it is also a vital aspect of a sustainable world society. The majority of people should be proud and content to conserve their local environment by living in tune with it and the resources it provides. To a great extent, this is how the world was half a century ago. In 1961, almost all countries in the world had more than enough capacity to meet their own demands (Hails 2008). The world must find once again the ability to balance demand and resources. It is unrealistic to believe that this can be done as a coordinated world community; changes must come from the bottom up with people being encouraged to make a success of the community around them.

Local Decisions and Local Circumstances

Population dynamics in a sustainable world will be based on local decisions and local circumstances. As governments take greater responsibility for striking the balance between economics, environment and social provision, there will be much more interest in defining membership of a particular society. With resources coming under pressure across the world and ensuring the security of supply becoming a policy imperative, a consequence will be that barriers to immigration will increase.

A crowded developed country, such as the UK with a population of 65 million, will have to operate a three-pronged policy. First, to reduce consumption as far as is feasible for a north European country. Second, to ensure secure supply routes for additional resources. Third, to introduce policy that, over time, reduces the population (OPT 2009).[3]

Each person in the UK has an average ecological footprint of 5.3 gha compared with an ecological capacity per person of 1.6 gha (Global Footprint Network 2008). Leading examples of low-impact sustainable communities in the UK such as the Beddington Zero Energy Development (BedZED) in London and the Findhorn Community in Scotland manage to live with ecological footprints of 3.2 gha and 2.6 gha respectively (Tinsley and George 2006). If 3.2 gha is taken as an achievable figure for the average ecological footprint of a UK

3 A poll, commissioned by the Optimum Population Trust for World Population Day (11 July 2009), found support for the view that both the world and the UK are overpopulated. Seventy-two per cent thought world population was too high, causing serious environmental problems, and 70 per cent took the same view of the UK.

resident, the maximum population that the UK can support sustainably from resources within its own borders is 30 million people.[4]

Populous countries that are consuming way above the local ecological capacity will be most exposed to the coming world resources squeeze. These countries will have to work hard to establish sustainable policy before the world hits the limit of available resources. International agencies and NGOs will have to understand sustainable policy choices and push for sustainable solutions in the advice they provide.

One difficulty is ageing population and the load that this places on the economy. Simple economic analysis leads to proposals such as increasing birth rate, or increasing immigration, to maintain a balance between work-age and retired people. We have to break out of this circular argument. The sustainable solution is to plan fiscal policy for a reducing population that has a high proportion of older people. It may take decades before the population bulge works through the system. This may be difficult, but it is a challenge that must be faced in order to be able to balance population with the availability of natural resources.

Balancing Population and Resources

At the heart of sustainable policy is a balance between population and resources. This truth is inconvenient, so it is often skirted around. This denial is dangerous. When world markets cannot deliver enough food and resources, there will be a scramble for supplies and considerable suffering amongst the poorest and least powerful nations.

When food is in short supply, and fuel to keep warm is scarce, communities will close in on themselves to try to survive. If this happens in a mild way on a country-by-country basis, it will force the pace towards sustainable policies. A certain amount of discomfort and inconvenient shortages will focus minds and change attitudes from the wasteful ways of the last three decades.

By continuing to rely on an open world market, shortages will not bite as soon and the crisis over resources will be delayed. A shortage in one place can be compensated for by importing excess capacity from somewhere else. When

4 The total ecological capacity of the UK is 96 million gha. With an average consumption of 3.2 gha, the sustainable population would be 30 million people.

the crunch comes, and the whole world market is under pressure, prices could increase rapidly and severe problems result. As soon as this looks likely, the reaction of governments, corporations and consumers will be to hoard supplies, making the crisis much deeper, longer and harder to counter. The ban on rice exports, implemented by India in 2008 in response to global food shortages, is an example of what is to come.

In places where population and resources are so out of balance that people are struggling to survive, extreme measures will have to be considered. Popular opinion may then support policy that, in a more stable society, would be regarded as unfair and unacceptable. The circumstances could allow far- right extremists take control. When one's family is struggling to make ends meet, the bigoted policies of such organizations may seem attractive. It could be far worse. I worry that, unless we can improve the population dynamics of the world, the consequence will be the emergence of leaders who exploit the coming crisis with genocide in mind. When people are starving, the extermination of another tribe or another ethnic group may seem a reasonable policy from the perspective of those who are suffering.

Improving the world's population dynamics is another consequence of pursuing sustainable policies. The world needs a mosaic of different societies and cultures, each of which balances the population with its resource base. Drawing on resources from elsewhere is a risk. Countries will have to look carefully at the sustainability of the source and be robust in defending supply routes, either through paying high prices or physical protection, or a combination of both.

The most sustainable and lowest-risk policy is to discourage further population growth. In this way, communities can draw back to within the capacity of the locality. Development based on additional consumption will have to be connected with further reductions in population. Most places on the planet can be a paradise for a population of the right size, but getting there will be exceedingly hard.

9

Finance and Capital Flows

Such global economic interdependence has not been seen before. Is this a strength or a weakness? Is the world financial system a robust self-regulating system or a house of cards waiting to collapse? The amorphous nature of the system makes it hard to judge. It is certainly looking like a system in which we all either stand or fall together. It may be that more connections bring greater resilience and reduce the chances of collapse, but if collapse does come, there will be no hiding from the consequences.

I wrote the preceding paragraph in early 2007 before the financial crisis of 2008 (McManners 2008: 171–2). I was not to know how prophetic my words would be. The context of the paragraph was part of my analysis of the role of capital markets in a sustainable society. My general conclusion was that 'capital markets ... have developed to suit a narrow economic focus ... [and] ... the relentless drive to deliver short-term results is not conducive to building a sustainable future'. I made a number of proposals for improvement.

Following the financial crisis, there is a danger that, instead of improving the system, blame is attributed to the specific triggers that initiated the crash. This would be dangerous. Such an approach may lead to creating safeguards against a recurrence of the same set of circumstances, instead of addressing the core problems at the heart of global finance.

Global Finance

Global finance has become more complex and more interconnected than at any time in history. Over the last three decades, there has been a huge increase in international capital mobility. In 1975, the market was driven by real needs:

about 80 per cent of foreign exchange transactions were to conduct business in the real economy to produce or trade goods and services. The remaining 20 per cent of transactions were speculative. By 1997, the balance had shifted dramatically. Bernard Lietaer (1997) estimated that just 2.5 per cent of foreign exchange transactions in 1997 were in support of the real economy. Now, the overwhelming majority of currency trades are for the sole purpose of profiting from buying and selling currencies. Other cross-border financial transactions have also increased dramatically. For example, gross cross-border transactions in bond and equity for the United States increased from 4 per cent of GDP in 1975 to 100 per cent in the early 1990s and grew to 245 per cent by 2000 (Hau and Rey 2006).

The bulk of world capital flows are pure financial transactions by currency traders, speculators and other investors shifting capital around to achieve the greatest return. The complex mesh of international finance makes it hard to judge whether this system is stable or not. The complexity hides risks that may only be exposed in a crisis.

The world financial system has evolved from a traditional model of banking to a commercial business focusing on massaging complex deals to conjure up profits. The financial markets have become self-serving and, in some respects, little more than gambling dens. The time has come to return finance and the financial markets to their true purpose.

Finance needs to be simplified – in order to be able to understand and manage these risks – and demoted from being the prime aim of policy to act in support of sustainable policies. Finance is of no value to human society unless it is used to deliver improvements in people's lives.

A return to transparent banking is needed so that it can act in support of the real economy to help people with the transactions of life, free from unsustainable and irresponsible speculation.

The Fundamentals of Finance

The fundamental reason for money is to facilitate the exchange of goods and services. Instead of bartering, which requires matching one object to be traded for another, the deal is agreed for money. This then moves around and is exchanged for other goods and services. At first, money had real intrinsic value, for example, gold coins. The next innovation was to lock the gold away in bank vaults and print paper promising to pay the bearer on demand the sum of whatever denomination was printed on the note. In theory, the holder of the paper money could take it to the bank and exchange it for gold. In reality, the paper money was accepted as a token for exchange and the gold stayed locked away. Central banks then realized that they had built up such trust in their currencies that they no longer needed to back it by gold. The UK dropped the gold standard during the Great Depression of 1931. The United States dropped the gold standard in 1971 at the time of the Vietnam War when US foreign debt was growing. All major currencies have followed suit.

In addition to acting as a token to exchange goods and services, money has a role in making capital projects possible. To construct a building requires upfront purchase of the materials and labour to build it. The payback in rent may take place over a period of decades. Banks therefore package up the cash they hold on behalf of savers into loans for projects such as house building. The borrower then pays back the loan over a set period including interest. The saver receives interest at a lower rate and the bank covers its expenses by keeping the margin.

The system of savings and loans is totally reliant on trust. The banker holds the savings of a number of people and a loan book. The borrowers pay monthly interest and the savers receive monthly interest. The bank will hold very little cash: just enough to cover the normal expected level of withdrawals. With a lot of borrowers and a lot of savers, on average the bank's books balance, but the bank no longer has the cash that savers have paid in.

This simple model of banking is easy to understand and the risks are transparent. If a rumour circulates that a bank is in trouble, this may prompt savers to move their money. Once the bank's reserve of cash has been paid out, the bank needs to raise more cash. It cannot reclaim its loans immediately; it has to look to other banks for funds. Other banks then look very closely at the bank in trouble. If other banks believe that the troubled bank is well run and its loans are sound, they will lend (for a fee) and the bank is secure. The crisis is soon diffused. If other banks are not persuaded that the bank is sound and are, therefore, not prepared to lend funds, the bank cannot honour its obligations and goes bust. Administrators are then brought in to go through the books and unwind the financial affairs of the bank. Simple transparent banking operated by many separate banks within the single authority of one country is naturally self-regulating.

The Rise of Global Banking

The policies of the last few decades have encouraged deregulation and freed up flows of capital. This has allowed the growth of huge global banks such as HSBC and JPMorgan.[1] HSBC is listed in London and JPMorgan on the New York Stock Exchange, but their operations extend across the world, reaching into almost every country. In this interconnected financial mesh, capital shifts around the system looking for the best returns. Prior to the credit crunch of 2008, there was a huge pool of liquid capital to draw upon. The interconnections brought more apparent stability to each bank, and to each country, as local weakness could be countered by drawing on global sources of funds.

During the ten years that Gordon Brown served as the UK Chancellor of the Exchequer (1997–2007), his mantra was that there would be no return to 'boom and bust' economics. This reflected the prevailing view of leaders of world finance that the global financial system was basically sound. Gordon Brown believed that his national economic policy was prudent and that the UK would be insulated from the insecurities, volatility and instability within the global marketplace.[2] World leaders and policy makers were not aware that an international financial edifice had been built with the appearance of stability but which was structured like a house of cards. Global finance became so big and interconnected that a loss of trust could infect the whole system. Those people who should have understood the dangers, such as the central bankers, have a responsibility to maintain trust in the system so were wary of pointing out the weaknesses in fear that it might implode.

A single isolated bank would fold if it lost trust and all the savers tried to draw out their funds. The increased connection between banks has made the collapse of any one bank less likely but the bond of trust that that an isolated bank needs for survival now applies to the whole system.

The complexity of the financial system is now so great that how it operates is opaque to regulators and the banks themselves. For regulators, it is hard to assess whether a bank is sound. For banks, it is hard to judge whether another bank is a safe counterparty for inter-bank loans or the purchase of financial

1 The world's two largest banks by market capitalization as at June 2009 were HSBC and JPMorgan Chase and Co.

2 Gordon Brown, in a speech to the British Chamber of Commerce national conference, 5 April 2000: 'In a global marketplace with its increased insecurities and indeed often volatility and instability ... My vision is of a Britain where there is not stop go and boom bust but economic stability.'

derivatives. The big banks themselves have such a complex array of activities that chief executives have little chance of knowing the extent of the exposure and level of risk within their own bank.

Many banks have become so big that governments dare not let them fail, adding to the apparent resilience of the system. While economies were strong, the big international banks made good profits for their shareholders. When the economic cycle turned down, risky positions were exposed and 'being-too-big-to-fail' meant the government had to step in with support. If governments understood this implicit bond they would be very careful to regulate closely the big banks.

Governments are careful in their dealings with big international banks, because they are wary of the power the banks wield. The big international banks transcend national boundaries and so do not answer to any one government, although operations must conform to the financial authorities of the countries within which they do business. These banks have choices about where to locate their operations. Governments do not want to drive them away through regulation that is more onerous than in other financial centres. Governments go further than this, and compete to attract financial services by minimizing regulation. The 'light touch' regulatory regime of the City of London is one reason why London has grown into a major hub of global finance.

Complex, interconnected global finance, operating under minimum levels of government oversight, delivered outstanding performance in the financial sector for almost two decades until 2008. The presumption was made that the system was self-regulating in much the same way that a collection of small national banks operating a simple banking model is self-regulating. It took the financial crisis of 2008 to expose this assumption as false.

The Financial Crisis of 2008

Up to late 2007, huge profits were being made on a range of ever more complex financial derivatives traded on the international markets. These had become so exotic that few traders or bankers, beyond the financial boffins who dreamed them up, understood the detail of how they operated. No one understood the effect of so many derivatives interacting in a complex global market. The financial crisis of 2008 exposed the problem. The US subprime mortgage market is often cited as the prime culprit, but this was the trigger rather than the sole cause.

The financial crisis of 2008 will be pored over for decades as economists look for explanations and governments attempt to introduce regulations to prevent it happening again. We could end up with a yet more complicated system and hugely complex procedures of oversight. With a system of such complexity, unintended consequences would be inevitable. The situation would then be set for a repeat crisis initiated by a different trigger.

It is worth looking at the subprime mortgage market because of the insight it gives into the bigger issue of how the system of international finance, as currently set up, allows irrational and unsustainable activities to thrive. The US mortgage market became part of a complex web of international finance. Clever people designed a sophisticated system that was profitable (for the people running it), legal and also bound to fail eventually to the detriment of savers, borrowers and society. When the subprime market collapsed, the contagion spread rapidly throughout the international system to every economy, affecting every country.

Leading up to the collapse, housing mortgages were being arranged by brokers for people with no income and no assets. The loans were often greater than 100 per cent so that the borrower could purchase furnishings and be able to pay the early instalments. This was not just a US phenomenon – one of the UK banks destined to be rescued by the Government during the crisis (Northern Rock) was advertising 125 per cent mortgages. A traditional local bank would not have taken such a loan on to its books. This would have been regarded as incompetent lending and the manager responsible would have been reprimanded. However, a new financial product had been developed called a Collateralized Debt Obligation (CDO). These consisted of a package of mortgages sold off to institutions in chunks. The initial mortgage broker could pocket a fee and pass on the loan elsewhere in the financial system. In a market of rising house prices, it was assumed that, if the borrower defaulted, the property could be sold to pay back the loan. CDOs were, therefore, seen as safe. Selling CDOs as safe investments secured on property encouraged yet more subprime lending and pushed house prices even higher. This was a pyramid scheme that was bound to collapse eventually.

Looking back, it is fairly easy to see that collapse was inevitable, but the system was so complex (more complex than described here) and there were so many interdependencies that no one could predict accurately what might happen.

Compare this with a simple banking model. The annual accounts of my local building society, the Newbury Building Society, for 2008 show that it held £500 million of people's savings and had lent £506 million as mortgages. The difference is just £6 million and this was covered by £34 million of capital. In this building society, each mortgage borrower is carefully assessed and strict multiples-of-income rules apply. Priority for mortgages is given to those who have been savers for some time. This is finance run for the purpose of supporting local society. People save a deposit with the building society to then be able to take a loan to buy a house. Over the next 25 years, the loan is paid back. Older customers then tend to be net savers. The financial operations of the building society are transparent, low risk and provide a service to the community. In addition, the building society has remained mutually owned so there are no pressures from shareholders to squeeze out the maximum financial return. The management of the building society answers to the members – who are the savers and the borrowers.

I do not argue that we should regress to some quaint banking system from the past. Technology has moved on in leaps and bounds, but its capabilities to automate transactions and transfer funds across networks has to be according to policy, values and priorities set by people for the benefit of society.

Improving the World Financial System

If policy makers stepped back from the detail, there is a simple objective that could solve the weakness in global finance, though implementation may be far from simple. The objective should be to return world finance to its true purpose of supporting the real economy under close national oversight. Real transactions that relate to investment and the sale of goods can be transparent and easily understood – unlike complex derivatives traded in international cyberspace. Governments should encourage domestic banking to be carried out by national banks to ensure that there are clear lines of accountability to financial authorities. Discussion should also take place over whether government should own banks. Government ownership would ensure that they are run for the benefit of society, but this must be balanced with the risk of inefficiency when private-sector commercial discipline is removed.

Taking banks into public ownership would have been politically impossible before the financial crisis. During the crisis of 2008, the majority ownership of a number of banks, such as the British banks Northern Rock and Lloyds TSB, shifted by default to the Government as a result of being rescued from collapse. The UK Government says that it intends to return them to the private sector as soon as possible. The Government should review whether retaining state ownership, or a majority stake, could form part of a system that is less at risk of collapse and better able to serve society.

Governments may also support a new breed of mutual financial organization when they realize that the steady returns from simple 'boring' banking can serve society well over the long term.

The oversight required from national financial authorities can be minimized if banks are prevented from growing beyond a certain size. The system is not then at risk from the failure of any one bank. Governments can step back and, to a large extent, allow the banks to regulate themselves. In a transparent national system, bankers will be able to judge the soundness of other banks, which may result in occasional failures when banks take the commercial decision not to extend lines of credit to a suspect bank. Losses will be limited in scale and government-backed compensation schemes will able to cope.

International investment banks will continue but with limitations on the scope of their business, as countries that follow a national banking policy will resist foreign ownership of retail banking institutions.

There are other improvements needed, such as reducing the excessive role of speculation within financial markets, but first I will examine briefly the problem of imbalances within the global financial system.

Imbalances in the World Financial System

The underlying problem of international finance is that the world has an interconnected financial system but no true world financial authority or world central bank. A truly global financial system needs strong global financial institutions. In a world with separate economies and diverse centres of financial control, it is important that capital flows between economies come under the oversight of national financial authorities. It is dangerous to mix elements from these separate scenarios. Free flows of capital, without either powerful global financial institutions or being subject to close national oversight as capital crosses borders, is a system without effective controls. I argue that the direction the world should take is to strengthen the oversight of national financial authorities and insulate them from global dependencies.

The biggest global dependency in the world financial system is that between the United States and East Asia. In 2009, the United States had foreign debts that exceeded $4 trillion. In comparison, China held foreign currency reserves of approximately $2 trillion and Japan held $1 trillion. This situation has supported high levels of American consumption, allowing the United States to import far more goods than it exports. In exchange, it sells US treasury bonds. Such an imbalance in trade should lead to the strengthening of the Chinese currency, making Chinese exports less competitive and bringing trade back into balance. China keeps control of its exchange rate and has resisted calls to make a substantial revaluation of the yuan. There is mutual interest between these two countries to maintain the imbalance so that the United States can continue to consume and China can continue to grow its economy on the back of exports. Looking from a world perspective, this overhang in the world financial system is a distortion that could unwind with unpredictable consequences.

One of the reasons for huge levels of US debt is that the world needs a reserve currency and the currency of choice is the US dollar. The US economy has been the world's strongest economy so, in the absence of a better alternative, it was natural that the US dollar should assume this role. The demand for US dollars to be salted away in the reserves of other countries reinforces the US economy, allowing the United States to borrow at very low interest rates.

The United States is to blame, of course, for the high levels of spending that have increased the country's indebtedness, but it is hard for the country to avoid building up debt when other countries choose to use the US dollar as their reserve currency. A sustainable financial system needs to be in balance. A system that further reinforces an already strong economy does not seem sensible or fair. It is a problem for the United States, as it affects the competiveness of its own industrial base, as much as it is a problem for the world financial system. However, the world financial landscape is changing with the rise of another reserve currency, the euro.

The Rise of the Euro

Eleven European countries made up the euro area when the euro was introduced in 1999, with membership increasing to 16 countries by 2009. At its formation, there were some doubts whether such a diverse range of economies, from Germany to Italy, could fit inside a single currency. The doubters have been proved wrong – so far. The euro is a strong currency that is now a credible second reserve world currency. The euro countries benefit from the economic advantage of reserve status, but they also lose some control over economic decisions and may encounter the same problem as the United States of growing a mountain of debt. Having a second reserve currency is good for world financial stability, but it may not be good for Europe.

Demand for the euro leads to exchange rates with other currencies being higher than they would be otherwise, making exports outside the euro zone less competitive. Countries in the euro zone no longer have control over setting interest rates to suit their economies or their stage in the economic cycle. Some countries will suffer without the option to devalue their currency compared with other countries in the euro zone.

The main risk to the future of the euro is the lack of convergence between the national economies inside the euro zone. Differences in economic policy and performance lead to imbalances. Inside national currency areas, regional differences lead to the transfer of funds as regional aid. It is doubtful whether the rich members of the euro zone, such as Germany, would be willing to transfer significant wealth to the economically weaker countries on a long-term and continuing basis without closer alignment of fiscal policy.

For the euro to survive over the long term, Europe's economies need to continue to converge. It is possible that some European countries may yet backtrack out of the euro if the conflicts between national priorities and differing economic performance prove too great to handle. An alternative for Europe would be to have separate currencies with agreed cross-holdings between central banks to provide crisis reserves. This would reintroduce currency transaction charges for trade within Europe, adding a restraining influence on the free market. These negative factors would have to be balanced with the benefits of greater national control.

There is little prospect that the euro will be dismantled any time soon, but it is likely that one or more members will explore the possibility of exit to regain greater financial control. Weaker countries may have the decision forced upon them as imbalances within the euro zone increase. It is possible that one or more highly indebted countries suffers a loss of confidence by the financial markets and is forced to default on its sovereign debt. In such circumstances, it is hard to envisage how such a country could remain inside the euro. It is also possible that stronger countries consider exit for selfish political and economic reasons. Implementing the extraction of one country from the euro would be a challenge for both the country and the European Central Bank, but once one country had acted as trailblazer, others might follow.

International Monetary Reform

Greenwald and Stiglitz (2006) presented a proposal for international monetary reform to the American Economic Association which would reduce the role of the US dollar as the world's reserve currency. Their proposal was to make more use of Special Drawing Right (SDR), a reserve asset administered by the IMF.

The SDR is a potential claim on the currencies of IMF members that can be drawn upon in a crisis. SDRs were created in 1969 to support the Bretton Woods fixed-exchange-rate system. To maintain exchange rates, countries needed official reserves that could be used to purchase the domestic currency in world foreign exchange markets as required. But the international supply of two key reserve assets – gold and the US dollar – proved inadequate to support the expansion of world trade and financial development that was taking place. Therefore, the international community decided to create a new international reserve asset under the auspices of the IMF. As the major world currencies shifted to a floating exchange rate regime in the 1970s, the need for SDRs was reduced.

The Greenwald and Stiglitz proposal was that SDRs should be issued on a substantial and regular basis and credited to the IMF accounts of members. In this way, the world demand for reserves could be satisfied without increasing the demand for US dollars.

Joseph Stiglitz (2006) went further and proposed a new currency that he terms the 'global greenback'. John Maynard Keynes had already recognized this potential solution and called his money 'bancor'.[3] The basis of the concept

3 The bancor was a world currency unit proposed by John Maynard Keynes, as leader of the British delegation and chairman of the World Bank Commission, in the negotiations that established the Bretton Woods system, it but was never implemented.

is that the countries of the world agree to a new currency – let's call it 'world money' – that in times of crisis could be exchanged for their own currencies. A closer examination shows potential problems. As Stiglitz recognized, the new currency could become subject to exploitation by central banks taking advantage of currency rate movements. There would need to be controls of when, and in what circumstances, central banks could make conversions into and out of world money.

The idea of world money may yet gain ground but it perpetuates the concept of global finances and a closely coordinated world financial system. Currencies require a central bank to manage them. World money would require a world central bank. This would increase interconnection and interdependence when what is needed is quite the reverse. Economies need to be decoupled so that problems are isolated. When a crisis hits, the countries that are not affected would have the resources to provide monetary assistance orchestrated by organizations such as the IMF.

The IMF has an important role as the lender of last resort, but its prescriptive one-size-fits-all policy has not always been helpful. During the Asian financial crisis of 1997, the IMF imposed policies that caused considerable hardship to the countries concerned. The IMF had a particular, narrow view of what it regarded as the correct policies and it insisted on structural reforms cutting government expenditure, reducing deficits and allowing insolvent banks and institutions to fail. These policies were the opposite of Keynesianism, which would support increasing government expenditure and propping up major companies. The role of the IMF was widely criticized and generated considerable resentment against the IMF.

In a world community that reinforces the role of the state and supports freedom to choose local policy, expanding regional initiatives may be more beneficial than increasing the role of the IMF. It should be easier to set up arrangements between countries with close economic links, as well as shared values and compatible cultures. The ASEAN group of countries is an example. They have increased their monetary cooperation and agreed to cross holdings of each other's currencies. This provides the reserves they need without having to hold such large quantities of US dollars or euros.

Proximization and Financial Management

The policy framework of proximization, as put forward in this book, leads to considering divergence of economies and separation of currencies as the basis of safe and stable macro world economics. Countries should be free to set their own economic policy as one leg in their sustainable policy framework that suits their culture, geography and resources. Instead of further integration, we need separation. The economic penalty of backtracking on economic globalization is balanced by increased stability at the macro level and greater flexibility at the national level to adopt a sustainable policy framework.

Proximization leads to renewed support for national currencies and to consideration of the benefits of local currencies such as their role as a mechanism to increase employment. At its simplest, this is just one step further than bartering. An example might be a parent's club to provide babysitting services. Any member can ask any other member to babysit their children. A central register is required. Over the long term, everyone is expected to take and give in balance. Such coordinated bartering of services can be extended by issuing IOUs for any goods or services within the local community and logged within a central register. This register would be the equivalent of a central bank for IOUs. People can earn IOUs by helping others, or spend IOUs by engaging help. Such IOUs would represent a simple currency. A further enhancement would be to set an exchange rate between the community's IOUs and the official currency of the country. This would allow richer people or old people on pensions to use the services of the community in exchange for a real cash contribution. Other people who carried out many community activities could earn IOUs which they would be able to exchange for real money.

Local currencies have high potential to support sustainable communities through a stable local economy with some protection from external factors. The WIR[4] bank in Switzerland, which was founded during the Great Depression to counter currency shortages, is an example. It is a barter system between small- and medium-sized businesses with an annual turnover in excess of 1 billion Swiss francs. At this scale, WIR may have contributed to the remarkable stability of the Swiss economy.

4 WIR is a not-for-profit bank serving small- and medium-sized businesses. It exists as a bookkeeping system, with no script, to facilitate transactions. WIR was founded in 1934 by businessmen Werner Zimmermann and Paul Enz as a result of currency shortages after the stock market crash of 1929.

Local Exchange Trading Schemes (LETS) originated in Canada in the early 1980s operating with interest-free credit to facilitate the exchange of goods and services. There are now over 300 small not-for-profit LETS schemes in the UK (LETSlink UK 2009).

There are also examples of local paper currencies, such as the BerkShares (2009)[5] which have been used in the Berkshire region of Massachusetts since 2006, and the Lewes pound which was launched as an experiment in the English town of Lewes, Sussex in 2008.

Governments become concerned when local currencies and LETS reach a significant size. The latter are clearly beneficial to the community in which they operate, but they also screen economic activity from national tax authorities. From the perspective of the state, such local currencies formalize a black economy and reduce the taxes collected by the state. The state can either control such initiatives to limit their scope or alter the national taxation scheme to accommodate local barter/currencies. The potential of local currencies to support vibrant sustainable local communities is so great that this second course is worth considering.

There is a fundamental conflict between conventional taxation based on income and sales tax, and local barter/currencies. One approach would be to bring a local tax system into these local economies, but this has the risk of adding layers of bureaucracy and complexity. The better approach would be to change the tax system.

Chapter 5 discusses shifting the tax burden away from income and expenditure and on to resource extraction and land taxes. If this is implemented, the threat to government of local barter/currencies is much reduced. Taxes on the extraction of natural resources and a land tax are both sources of revenue that are unaffected by systems of local bartering for goods and services.

If green taxes are taken to their logical conclusion, taxing income and work is not required. There is then total freedom to develop local barter/currency systems that provide very stable local economies, insulated not only from global instability but also – to a certain extent – from economic imbalances within the country.

5 BerkShares are a local currency designed for use in the Berkshire region of Massachusetts and issued by BerkShares Inc., a non-profit organization working in collaboration with participating local banks, businesses, and non-profit organizations.

In developed countries, it is unlikely that governments will want to dismantle complex tax systems that are accepted by the electorate and effective in raising revenue. However, many poor countries might find such green taxation a relatively easy route, considering that the de facto system in place may already be a large and pervasive black economy.

The connection between green taxation and sustainable monetary policy is another example of how different strands of the proximization policy framework reinforce one another.

Supporting Poorer Countries

This discussion about reforming the world financial system points to an alternative way forward. Economists and politicians will have to work on the detail. The central point is the need to change the system and ensure that it is stable, sustainable and works for the benefit of all countries. In the current system, the countries that do least well are the poorest countries. The richer countries with economic power have a moral responsibility to ensure that vulnerable countries are not unfairly penalized as they act in defence of their economies.

In an open global financial system, banks look for opportunities to lend wherever they can find borrowers. The theory of free markets and free flow of capital tells us that this is efficient economics. The theory holds up well for capital flows between sophisticated developed countries. It does not work well in practice for vulnerable poor countries.

Poor countries can come under considerable pressure to accept loans from international banks. Being short of funds, such countries are, of course, tempted to accept the offers. The lender will protect their position by insisting that the loan is repayable in a major currency such as the US dollar or euro, leaving the country receiving the loan with all the risk of currency fluctuations. If the country struggles to make repayments and its currency slides, the difficulty of paying can become extreme. Some very poor countries have been saddled with debts that require a considerable portion of meagre budgets to service the loans (Stiglitz 2006).

The IMF steps in to offer cash to repay the loans, but under strict conditions that may not suit the country. IMF loans are not debt forgiveness. Often the money goes directly from the IMF to the Western banks that made the loans. The banks are happy with such an arrangement, of course. The country is then responsible to pay back the loan to the IMF, a more senior creditor with considerable power.

During the 1990s, campaigning by a broad coalition of development NGOs, Christian organizations and others forced the issue of Third World debt on to the agenda of world leaders, leading to protests at the G8 meeting in Birmingham in England in 1998. The campaign was successful in persuading the IMF and the World Bank to set up the Heavily Indebted Poor Countries (HIPC) initiative to provide systematic debt relief for the poorest countries.

The danger is that debt relief then sets off another round of unsustainable lending. The international banks observe the actions of the IMF and believe that they will get their money back even if they lend to countries that have little chance of ever being able to pay the loan back from their own resources.

Such international lending, consisting of foisting loans on to countries that cannot afford the repayments, has close parallels with the subprime lending market. The lending banks are as much at fault as the borrowing country – if not more so, as the banks should understand their business and the risks involved. This may be profitable for the banks, but is irresponsible behaviour, and it can be prevented.

Poor indebted countries would often do better not to accept loans from the IMF, even if this meant defaulting on their debt to international banks. A better method than loans from the IMF of providing help to such countries would be to persuade international donors to make grants of aid earmarked for specific purposes to help the country improve. This would remove the moral hazard that afflicts the lending by international banks. Without the backstop of IMF lending, it is far more likely that poor indebted countries will default on their debt in times of financial crisis. The risk to the poor country is that this will make it harder to raise new loans in the future. But this may be a benefit, as borrowing too much on terms that suit the lender is one of the main problems they face. If a bank lends to a poor country, it should understand that, if the country cannot repay the debt, the country is likely to default on the loan.

Odious Debt

Exposing international banks to the true risks of the loans they make will also be useful in reducing levels of odious debt. This is national debt taken on by oppressive regimes to, for example, purchase weapons to keep their own people under control, or to enrich the personal accounts of the leaders. When finally the country returns to democratic rule, it may be saddled by huge debts. It is iniquitous that an oppressed people should have to pay back the costs of their oppression.

This situation was encountered by Iraq after the country was freed from the rule of Saddam Hussein. In 1927, Alexander Sack, a Russian international law scholar working in Paris, formalized the principle that such odious debt should not be enforceable under international law. Patricia Adams (1991) argues that such debts are owed by the regime that incurred them, not the state. Adherence to this principle would change the behaviour of international banks. Instead of lending to totalitarian regimes and turning a blind eye to where the cash goes, they would have to assess the risks very carefully. This would have a humanitarian benefit of making it very difficult for countries with a poor humanitarian record to secure loans. By declaring that any new debt will be regarded as odious, and therefore not a liability that the state has to repay, sources of funds would close down, acting as a strong complement to other sanctions (Jayachandran and Kremer 2006).

This approach would mean that decisions about international debt are based on real legitimate needs and a proper assessment of ability to pay, bringing more stability to international finance.

Reducing the Power of Speculators

Encouraging governments to back off from trusting a globalized financial system, take more national control over their banks and accept responsibility for their financial affairs would foster a level of prudence that should be normal. This may not be enough to confine international finance to a support role. The problem of excessive speculation also needs to be addressed. There will always be opportunities to profit through arbitrage between markets. When such trades are a small proportion of capital moving around the market, they serve to even out imbalances and keep the market liquid.

When the majority of money flowing through the financial market is speculative and an order of magnitude greater than that needed by the real economy, speculation becomes a powerful force. Instead of providing liquidity, such huge flows have the power to distort and undermine the market.

To ensure that finance serves society, rather than acts as a form of gambling, the scope for speculation and manipulation of markets must be reduced.

Globalization, leading to open and interconnected markets, has increased the opportunities for speculation. Globalization has also encouraged competition between markets, forcing down the cost of transactions. Looked at in isolation, this appears to bring greater efficiency. Lower transaction costs mean less economic value extracted from the market by taxes or fees. The United States does not tax share transactions and, in an open global market, other countries have tended to follow suit. In the UK, tax on share transactions, Stamp Duty Reserve Tax (SDRT), was reduced from 2 per cent to 1 per cent in 1984 and further cut to 0.5 per cent in 1986. Germany, Sweden and Finland abolished share transaction taxes in the early 1990s.

The reduction in trading costs has made speculation easier and more profitable. When transaction costs are low, very small differences in price between one market and another can be exploited profitably. The way to reduce the scope and power of speculators is to reintroduce resistance to trading through a tax on financial transactions. The small discrepancy between prices that will always occur in markets will only be profitable when it exceeds the value of the tax. The turnover will be markedly less and will be dominated by transactions that reflect real needs in the economy.

I argued for increased share transaction taxes to support corporate sustainable strategy in *Adapt and Thrive* (McManners 2008). Such taxes would provide some protection from the short-termism that is endemic in current equity markets and which acts as a barrier to making the long-term business decisions that sustainability requires.

Taxation can also be applied to reduce speculation in international currency transactions. James Tobin (1978) proposed such a tax to limit the role of the speculator. He went further, suggesting that the tax receipts could be held centrally to support some aspects of the work of the UN. This additional measure had intellectual appeal to people concerned about helping poorer countries and ensuring that the UN has the funds it needs. It was unfortunate that he raised this complication because it distracted discussion away from the core purpose of the proposal to reduce speculation in the currency markets.

My proposal is to encourage governments to increase tax on all financial transactions from currency trades to share transfers.

Implementation will cause pain in the financial markets, and those responsible for running them will object strongly. The increased cost of transactions will cause turnover in the market to plummet, with the tax going to the government. Unless transaction fees rise, those running the markets will lose income in direct proportion to the downturn in turnover, as well as having to contend with collecting tax and passing it on to government. Where a tax already exists, changing the rate will be straightforward. The UK Government already levies a 0.5 per cent tax on the London equity market: the administration of a higher rate would require little change to systems and processes.

For my proposal to work there must be clarity of purpose and simplicity of process. The purpose is to introduce friction into global financial markets to reduce the power of speculation. The process should be direct taxation by national governments on the markets they control without the further complication of an international collection mechanism.

A national government is not going to take a unilateral decision to increase tax on financial transactions without considering the consequences. One consequence will be that multinational corporations will shift their currency trading and share listing to the markets with lower taxation. Governments, therefore, have to consider two options. One is to attempt to design regulation that requires certain trades to be made on national markets. The other is to broker an international agreement that sets a floor tax on financial transactions.

An international agreement to set a minimum level for tax on financial transactions is such a fundamental reversal of policy that it is not likely to happen soon. It requires understanding, and then adopting, a substantially different approach. Despite the inertia that there will be against this proposal, the logic for taxing financial transactions is sound.

Opponents will claim that the tax will be a drain on the markets. The value 'lost' from each transaction, taken together, will be a brake on the economy. This is a valid viewpoint and is worth a closer examination. This shows that value is not lost but transferred from the economy to government income. The government can then feed this back into the economy by reducing other taxes or increasing government expenditure. Such a tax will change behaviour (for the better), but at the macro level it can be tax-neutral and need not be a brake on the overall economy.

Compare this with the current situation where markets are dominated by flows of speculative capital. Each successful speculative trade takes value out of the market to appear as profits of traders and hedge funds. It is very hard to quantify the overall impact. There are high-profile examples where figures have been calculated. For example, on Black Wednesday[6] 1992 the UK pound was forced to leave the European Exchange Rate Mechanism (ERM) by currency speculators. The most notable of these was George Soros who is reputed to have earned over US $1 billion as a result. A devaluation and exit might have been required eventually, but if the UK Government had controlled the process it would have saved the country over £3 billion (McManners 2008: 172). A tax on currency transactions would not totally eliminate the risk of such financial ambushes, but the opportunities would be far fewer.

Once systems are in place to tax financial transactions at a rate high enough to spike the speculators, a high proportion of financial trades will be the result of real needs within the economy. The difficulty for those countries leading this policy change will be ensuring that, during implementation, the standing of their own financial markets can be protected within the world financial system.

Bringing Finance Back Under Control

World finance has been revolutionized by open loosely regulated markets supported by advanced IT systems. The result is making it harder for countries to control their economies and this will be a barrier to implementing the transformation to sustainable societies. In a sustainable world, the disciplines of banking and prudent financial management must be brought back. The market makers and international banks may protest, but governments will be doing their duty in facing down criticism and bringing finance back under the control of national governments.

6 'Black Wednesday': 16 September 1992.

10

The Global Knowledge Economy

Modern civilization is enormously complex and has been built on an explosion of knowledge facilitated by advances in IT and information management. The quantities of data held on computer are ballooning and the power of computer processors is doubling every two years.[1] The global knowledge economy, facilitated by these computers and networks, will continue its rapid growth. Sharing information at the global level will be important in disseminating methods and technology for sustainable living. Explicit knowledge can flow across borders with ease, but much valuable know-how is complex and specific to the context in which it resides. It is, therefore, also important to support local centres of expertise in order to develop innovative solutions that are insulated from the assumptions and restrictions that come from 'global group think'.

Knowledge is far more than the data and formal records held in archives. Ever since Homo sapiens started their successful ascent to establish human civilization, knowledge has been retained, shared and passed on to the next generation. Facts were originally passed on through memorable stories and fables and are now available from a myriad of sources. But knowledge also encompasses skills, culture and attitudes that influence how society is managed and operates. We have become very good at managing hard facts with the support of our computers, but retaining 'soft' knowledge continues to require primarily human engagement and expertise.

The formal approach to knowledge management is to record facts and links these facts through sequences of logical deductions. Adherence to this approach is based on the assumption that every problem can be solved by a sequence of appropriate analysis. Researchers in academia are constrained by

1 Over the last 45 years, computer processing speeds have approximately doubled every two years, fulfilling the prediction made in 1965 by Intel co-founder Gordon Moore that the number of transistors on a chip will double about every two years. This has become widely known as Moore's Law.

the requirements of research methodology. Each research student studying for a doctorate spends a large proportion of their time examining what has already been discovered. This leads to formulating a hypothesis which is then tested by experiment. The results are collated and conclusions are drawn. Knowledge is then advanced by one small step. Without such methodology, research would be anarchy.

The methodical expansion of global knowledge also has its drawbacks. Ideas are established and perpetuated as each contributor seeks to expand the body of knowledge. The development of knowledge must also involve challenging established ideas, freedom of thought and an environment that allows ideas to be aired and tested. Theories are proposed, researched and adopted with ever more references to the original research. This process builds up considerable bodies of knowledge that may be rooted in a few key assumptions. Challenging existing knowledge is a vital safety valve. Without it, civilization could end up in a rut of atrophied thinking. Informal human systems, consisting of memory and verbal knowledge, naturally eliminate old ideas as they drop out of circulation. Computer databases can retain all their data unless we make the effort to remove data from the active system.

There sometimes comes a point when key assumptions that are holding together huge areas of perceived knowledge are no longer valid and are blocking progress. This is the stage the world has reached with an almost unquestioning commitment to free trade and open markets. The deep-seated assumptions underlying this commitment have to be challenged in order to open up research and start the dialogue for the next wave of innovation.

A brief example illustrates my point. In discussion with a colleague, the topic of Africa arose. We had both spent time on the continent and shared a high regard for the people of Africa and would like to support measures that addressed some of the problems. I attempted to steer the dialogue towards my ideas about sustainability as the way to support improvement in people's lives. I quickly ran into a problem. My colleague was an advocate of neo-liberalism and working on deep-rooted assumptions about the benefits of globalization. His concept of equity was based on the idea that everyone should have the opportunity to live a lifestyle that matched his own, on the implicit assumption that every society should be helped to follow the development path of the West, and that free trade, open markets and free flows of capital were the way to achieve this.

The world needs other concepts to replace the old concept of globalization but, until we accept that economic globalization is no longer the appropriate basis for human development, it is hard to build new structures of thought. We are forever trying to add refinements to an edifice that is starting to show cracks, when the action required is to underpin our thinking with new foundations.

My concept of proximization can fulfil this need. As the basis of policy making, it will bring a new structure and new dynamics to how the world is viewed and how problems are solved. The theory that economic globalization is the best policy for the world can then shift from current policy to the historic record. In order to be able to react to changing circumstances in this responsive manner, we need a global knowledge economy that is able to handle ambiguity and which encourages the development and growth of ideas.

The Nature of Knowledge

Explicit knowledge is easy to store in written documents or databases, easy to share through libraries or online repositories and easy to transfer to people anywhere in the world at the click of a button. The facts available include measures of economic progress such as GDP and measures of human welfare based on health statistics such as infant mortality. Of course, the knowledge we need to run human society goes much deeper than this. We also need knowledge of culture and attitudes, human expectations and how to manage them, and an understanding of the nature of human welfare and happiness. These are complex issues that are hard to define.

Even knowledge that appears to be explicit often is not. The manufacture of computer chips is an example. There is a huge quantity of literature and designs that explains how this is done. It is easy to forget that these records alone are not enough to be able to make computer chips. We also need a society that delivers two solid decades of education to scientists so that they are able to understand the records. Technicians need decades of hand-on experience of how to make the complex systems work. High-technology manufacturing facilities are dependent on an extensive network of companies and experts who have dedicated their working life to filling their niche in industrial society. The information stored in databases is only a part of the picture. The knowledge required to make computer chips is a complex collaboration between people and machines.

The global knowledge economy needs to develop and protect these complex centres of knowledge. There is one category of knowledge that the world needs to expand urgently, and that is the methods, processes and technology needed to support a sustainable human society which is in tune with nature and protects the ecosystem. This primary requirement should direct the powerful engine of the global knowledge economy.

The growth in knowledge is partly a consequence of humankind's innate inquisitiveness but other forces also come into play. Times of difficulty and hardship can bring out the best in human ingenuity. The Second World War was a period of tremendous innovation and development, including radar to support the RAF in defending Britain's shores, the world's first computers to crack German secret codes, and the dubious benefit of nuclear weapons to bring the war in the Pacific to a close. The world may now be entering such a period of hardship that it will force the pace of change in developing sustainable solutions. It is unfortunate that severe hardship may be required to force the pace of innovation. Fortunately, there is another force accelerating the pace of knowledge, and that is commercial advantage and profit.

Twenty-First Century Comparative Advantage

Knowledge and innovation have always been important to the success of the developed nations. As globalization has worked to configure an efficient global economy (efficient from a narrow economic perspective), low-skill production has shifted to countries with low labour costs. The developed nations have responded by moving up the value chain to advanced products and high-tech solutions. Innovation is the key to success, and patent protection is vital to retain the value of research and development.

The sustainable world required for the twenty-first century is different. The physical production, consumption and recycling of products becomes a local business. Comparative advantage is not to be found in the production of physical goods but in the knowledge, design and know-how that is required. The importance of the global knowledge economy will increase substantially.

Technical knowledge of processes and designs is valuable and easy to send around the world. It is also easy to leak if care is not taken to safeguard sensitive research and technical advances. Infringement of copyright and the 'borrowing' or copying of designs is common in countries with immature or weak governance procedures. The future success of multinational corporations could hinge on their ability to generate and manage knowledge and convert this into designs and processes that can be manufactured in local production systems.

There will be a dilemma of safeguarding commercial success and rolling out sustainable solutions throughout the world. As the forces of twenty-first-century comparative advantage shape the world, corporations will search for knowledge to dominate areas of business. Countries will seek to establish world-leading knowledge hubs. At the same time, it is important that green technologies are rolled out rapidly, not just in the developed world but also in the developing world. Less-developed countries have the theoretical option to leapfrog the Western nations and bounce past our 'dirty' industrial phase. This requires reconciling the conflict between commercial interest and global environmental need.

Sharing the processes and designs required for sustainable living across the countries of the world is the obvious solution, but not if it undermines the forces driving research and development. Without investment for development, new greener technologies will not exist, and there will be nothing to share.

Distinctive National Knowledge Capabilities

The world needs distinctive knowledge centres capable of solving the challenges that we face. Some knowledge centres may develop the capabilities to take humans off planet Earth to explore beyond the solar system and occupy other planets. Other knowledge centres may develop the expertise to manage and conserve the Earth's ecosystems. Taken together, human society will gather the knowledge it needs. This will not happen if there is a push for global uniformity that stifles innovation. The world needs variety, and the knowledge economy is no exception.

Innovation is the key to success. This requires freedom to think and a willingness to challenge accepted logic. Wild ideas must be allowed to surface, be discussed and, if there is a glimmer of possibility that they might work, be tested. In the complex modern world, the innovations with the most potential may involve very complex systematic change that goes right across society. Sustainable societies will require the adoption of different bundles of policies, processes, technologies and methods. To test such radical potential solutions, experiments must take place on whole communities, or whole countries where the electorate can be persuaded of the potential benefits.

These islands of alternative ideas should include government policy, university research, business and society. A degree of isolation will be important. Those involved with innovation in business know well that big corporations can struggle to innovate. Corporate 'group think' can be hard to counter. Instead, corporations have been successful at hiving off particular projects, which are separated physically from the corporation and given considerable autonomy. This encourages the entrepreneurial free thinking needed to develop innovative products and different business models. Countries need to do the same to avoid global 'group think' taking control.

Commercial Research and Development

The profit motive is a powerful incentive to drive commercial research and development. This leads to focusing on solving the problems of the rich countries. A drug to counter the effects of obesity is likely to find a large market amongst affluent people. A drug to counter malaria would have a massive impact in improving lives in tropical countries, but these societies are poor and less able to pay. Commercial considerations lead to investment in obesity drugs rather than anti-malaria drugs (though this would change if, due to global warming, mosquitoes infected with malaria migrated north to afflict the rich countries of Europe and North America).

Commercial research is highly capable of developing solutions, but a way must be found to align a sound business case with the needs of world society and the planet. Before moving on to how to implement the sharing of processes and technology with poorer countries it is necessary to ensure that the general business case for research and development is sound.

The world needs universal and robust protection for intellectual property rights to ensure that the commercial incentive remains strong to discover and develop technologies and processes to improve the world. The application of the IPR will predominantly be licensing for local production – as this is how sustainable manufacturing will operate.

Corporations will be careful to keep close control of some key core technologies, such as computer chips, advanced robots and machine tools.

For example, a vehicle may be produced locally but a small central electronic controller may be needed that is only available from the corporation that holds the design and technology IPR.

We need to inject the concept of 'knowledge as aid' into this model of commercial research and development to ensure that the most sustainable and green technologies migrate quickly across the planet.

Knowledge as Aid

The green technologies we need to reduce humankind's impact on the planet and make the transformation to a sustainable society must not be restricted until the patent expires. These technologies must be rolled out quickly worldwide. In affluent countries, government regulation can drive the process of adoption and the corporations with the technology will profit from the sales. In poor countries, the environmental imperative to clean up processes and energy systems and build sustainable societies is just as strong, but the ability to pay is lacking.

One way to respond, that must be resisted, would be to divert aid money to purchase the licences required. The income of the corporations would be maintained (and with it the incentive for further developments) and the poor countries would get the technology. To get better value for the aid, it could be argued that the country should be persuaded to become an exporter to earn revenue from their new capability. This would then infuriate the corporation who owned the IPR, and who had been persuaded to sell licences at a discount only to find cheaper versions of the product being hawked around the world and threatening to undercut their business.

Diverting aid money to purchase licences for green technology makes little sense if the result is to perpetuate the old model of development based on industrial capacity geared to the export market. This would not only be a move away from the concept of sustainable local manufacturing, but also, on examination, it would help no one. The country would have less aid for other purposes, and the corporation owning the IPR would lose commercial advantage, so that further transfers of technology would be resisted.

My proposal is that rich countries and multinational corporations are persuaded to provide free licences to poor countries for sustainable technologies, but not as a charitable donation. The free licences would come with strings attached. These would consist of binding restrictions on the onward export of goods manufactured under the licences, as is common in commercial licensing contracts. The country would be free to build up manufacturing for its own society but would have to accept a ban on the relevant exports. If a country in receipt of knowledge aid decides to become an exporter, using the free intellectual property, the IPR owner would be able to make a claim for compensation at full commercial rates.

The country receives the technology it needs and the corporation's IPR is protected. Such low-income national markets are not attractive to multinational corporations because of the society's lack of ability to pay. Free technology is enormous assistance to the country at little cost to the corporation. In these circumstances, the corporation may act solely on the basis of improving its reputation within affluent markets.

There may be complaints that my proposal seeks to hold back poor countries from the possibilities of export-led growth and that it is contrary to the advice from, for example, the World Bank, which supports building Western-style economies. My approach is different. Poor countries should invest in building a sustainable local society by fully exploiting the opportunities of knowledge aid to improve the lives of their population. There is no need to enter a competition with the world's industrialized countries.

Competition makes free trade and open markets effective in delivering macroeconomic outcomes, but it comes with penalties, particularly for poorer countries. In championing free trade, the West is confident of maintaining a competitive edge in certain sectors, such as high technology or financial services. The West will defend this lead vigorously because, in an open world economy, each country has to have an area of expertise at which it excels and which is the basis of its competitive advantage. The West will do all it can to ensure that it is not overtaken by poorer countries that have been encouraged to follow a Western development path. This is not a deliberate prejudice against countries that are currently poor; it is simply the tough discipline of prospering in a competitive world. Countries that are currently poor and which follow advice from the World Bank to build industrial capabilities to serve an export market are destined to become second-class Western economies. The alternative for poorer nations is to aim to be first-class African, South American or Asian economies.

Countries that are currently classified as poor have the potential to become vibrant examples of good living through free technical designs and developing expertise in other ways, such as sustainable harvesting of rainforest products that keeps the forest intact. Such countries may not rise up the industrial production league tables or match Western levels of GDP, but they may build some of the world's most sustainable societies, well balanced and in tune with the ecosystem. Such societies could be the envy of the world. This aspiration is very different to the concept of industrial development widely promoted as the way to alleviate poverty whilst the theory of economic globalization holds sway.

The advanced Western nations can continue to dominate technological progress but our policy makers will have to visit the poorer countries to learn how to run a sustainable society. This is already true in many cases, but we in the West are often blinkered in what we see. When looking at poor agrarian societies, Western eyes see deprivation and a lack of material possessions. They see a problem to fix – to make the countries more like our own – rather than a different way of life with room for improvement.

Knowledge aid can work both ways. The West can share appropriate technology with the less-developed countries, and many less-developed nations can offer us advice on how to live with nature. In the global knowledge economy that I foresee, each country can lead in their own area of expertise, profiting where it is commercially viable and sharing expertise with less capable nations without risk. A reinvigorated knowledge economy is one aspect of a more effective model of global cooperation and coordination.

PART 3
A Changing World

11

Global Cooperation and Coordination

The challenges we face today are global in nature. By working together we can solve them.

Ban Ki-moon (2009)

The degree and complexity of the coming changes in human society is mind-boggling. This is not a situation that responds well to detailed top-down planning. The framework should be the minimum necessary to allow a balance to emerge between social need and respect for the environment whilst running stable economies. The detail has to be worked out within communities where there is the commitment to act and the willingness to compromise to reach solutions. The biggest effective community is the nation state. The fundamental basis of effective world policy must, therefore, adhere to the principle of subsidiarity. Global coordination and control should apply only in areas where solutions cannot be found at national level or where global coordination is more effective than national action.

Facing Up to the Challenge

The biggest problems humanity faces are global in nature and need higher level coordination. I put forward two risks capable of causing severe, long-lasting damage to humanity:

1. misuse of nuclear weapons
2. global ecosystem collapse.

The danger of nuclear proliferation threatens every country and every society. The acute dangers of nuclear technology have been evident ever since the nuclear bomb was invented and there is no dispute that coordinated action is vital. Considerable diplomatic effort is expended to agree and implement controls. I trust that the main nuclear powers will continue to work hard to ensure that the world does not face nuclear Armageddon. However, as less stable regimes acquire nuclear weapons the risks increase, not only of a state-sponsored nuclear war but also leakage of nuclear-weapon capabilities to terrorists. This is not a point I will dwell on here, except to point out that widespread adoption of nuclear power inevitably makes nuclear-weapon technology more available. Those who campaign for nuclear power to reduce carbon emissions need to be aware of this dilemma and ensure that the risks are factored in to any decision.[1]

The biggest immediate risk to humanity, nuclear war, is easy to understand and the world is taking substantive action. The other major risk to humanity is global ecosystem collapse. This is not such an obvious threat, but it could become the most serious if it is allowed to fester. It is a problem that the world is only now beginning to understand fully. The world is very slow to accept that there is an environmental crisis and even slower to commit to doing something about it. The agenda for change in world governance must focus on averting the pending world environmental crisis before the damage inflicted is severe and irreversible.

The threat of global ecosystem collapse arises from chronic imbalances between society and the ecosystem. At small scale, the impact is hardly noticeable but, as every country adopts similar consumption patterns, the potential negative impact is huge. The world community is slow to understand the true significance of this creeping threat and slow to devise global solutions. It is imperative that the world acts to fill this policy void. The need for global cooperation is greater than at any previous point in history.

The world is not approaching the 'dawning of the age of perfect global democracy' as Woodin and Lucas (2004) suggest. However world leaders, even

1 The debate over nuclear power leads to a divisive dialogue at world level where nuclear power may be acceptable in some countries but not in others, based on a judgement by one country over another. Western countries could choose nuclear power to delay fundamental change to the energy-usage patterns of their societies, but it would be hypocritical to stop nuclear power development in other countries based on a Western view of the extent to which they can be trusted. The nuclear debate is important and, as discussions continue, the risks must be taken fully into account.

those acting on a narrow perception of national self-interest, cannot ignore the growing calls for change. Michael Woodin argued for a Programme of Obstructive Deconstruction by governments to force change through disabling the institutions of economic globalization. The most powerful countries would be wise to neuter the potential of such action by leading change along the lines outlined below. Transnational corporations, accused by Woodin of having too much political influence, would also profit from understanding the nature of the coming changes and, rather than resist, help to orchestrate the changes (Chapter 13).

The current network of global organizations – such as the United Nations, WTO, IMF and World Bank – has achieved little progress in alleviating threats to the global ecosystem. The crux of the problem is that it is only the UN (through the UNEP) that has an explicit role with regard to the environment. The environment must shift from the margins of policy making to the centre of world power. This will not be easy. Decision making in global forums is often flawed as national representatives push selfish national priorities. Despite this, it is vital that world society agrees and implements an effective core framework.

The world needs a review as fundamental as the Bretton Woods agreements that shaped world institutions after the Second World War. Negotiations took place during a time of acute worldwide crisis and agreement was achieved quickly. The crisis the world faces now is severe, and the need for action is urgent, but the pace of progress is glacial. Humans respond well to acute crisis, but it seems difficult to inject the necessary urgency into the current crisis.

The image of world leaders that comes to mind is not a kind one. There is an oft-quoted experiment that describes dropping a frog into a glass of hot water. It will naturally jump straight out again. If the frog is placed in a glass of cool water, which is then slowly heated up, it will remain there slowly losing consciousness until it is boiled alive. Our world leaders are like the frog, happy to accept the current policy framework despite growing evidence of the need to make the leap to something radically different.

World leaders should find the courage to take the risk that their national interests may be compromised and convene a meeting of world experts, giving them the freedom to propose a new framework, unconstrained by policy of the past or vested interests in the current system. This chapter generates ideas to consider.

A Crisis of Legitimacy

World governance is weak as international bodies suffer from a crisis of legitimacy. The UN, established in 1945, is the most senior coordination and control body. There is the potential for the UN to become the overall umbrella organization for world governance, but this is not the current situation. Other key global organizations, such as the IMF, WTO and World Bank, are largely autonomous and report to their own boards of stakeholders.

The UN has limited power and is not in any sense a global government. Decision making is slow and consensual. In building a consensus, national representatives often push selfish national priorities. The UN has the Security Council to support the ability to react quickly in a crisis, but its membership reflects the world of the 1940s. Countries that will be important in the twenty-first century, such as Brazil and India, are not at the centre of power and feel marginalized. Lack of support for the UN extends to the United States, which has a long history of resisting the UN and withholding funds to seek to exert greater influence.

World financial governance is provided by the IMF and World Bank.[2] As a source of funds, these organizations have considerable direct power to influence countries by attaching prescriptive policy requirements to loans. The stakeholders behind the IMF and World Bank, in particular the United States but also other northern countries, have retained close control because they have not wanted to risk undermining sound economic management by ceding control. Despite such close oversight, in 2008 world finance suffered the worst financial crisis since the Second World War. It is widely recognized that the world financial system is in need of reform, but current effort is going into papering over the cracks rather than the deep-rooted reforms discussed in Chapter 9.

The organization that is most in need of reform is the WTO. The loss of perceived legitimacy arises from concerns in developing countries that the main beneficiaries of free trade are the most powerful nations and the global corporations. This is despite the fact that one of the WTO's principles is that trade should be more beneficial for less-developed countries. The problem is

2 Another component of the international financial system is the Financial Stability Board (FSB), established in 2009 as successor to the Financial Stability Forum (FSF) dating from 1999. This consists of central bankers and the heads of financial authorities with the mandate to promote stability in the international financial system.

deep-seated; the core principles that the WTO upholds are now out of date. The economic advantages of free trade are diametrically opposed to the consequential environmental impacts (see Chapter 7). The current institutional structure tends to divorce trade decisions from environmental considerations.

Adherence to the WTO policy package is no longer appropriate. While economic outcomes measured by GDP and quantities of trade were used as the yardstick of success, free trade was appropriate policy and delivered what was expected. The world has become a more enlightened place and a different set of priorities is required. Trade is a means, not an end. Trade should be subservient to environmental and social outcomes as discussed further below.

The UNEP is the most important international organization with regard to the world's current predicament. It was set up following the United Nations Conference on the Human Environment held in Stockholm in 1972. The work of the UNEP has defined the problems clearly and comprehensively, but it does not have power. The UNEP seeks to persuade and influence. This requires that those with power in world society are willing to listen.

Making the Changes

The International Forum on Globalization (IFG) has examined the weaknesses of the current system and recommends unifying global governance under a restructured UN (Cavanagh and Mander 2004). The IFG recommends closing down the IMF, World Bank and WTO and setting up a range of new institutions under the auspices of the UN. The IFG proposal further recommends that the UN Conference on Trade and Development (UNCTAD) becomes the primary UN rule-making body for international trade to ensure that there is bias in favour of the needs of low-income countries. The IFG proposal has some merits, but it relies on an idealistic view of the world.

Since its creation in 1964, UNCTAD has become the forum for developing nations to push for changes to the global economy to support their national development efforts. As the power and influence of the IMF, World Bank and, from 1995, the WTO increased, UNCTAD became increasingly marginalized. There is now a tension between the institutions in which the developed nations, principally the United States, have sway and UNCTAD, which acts as the voice of the southern countries. The idea that this balance of power could be overturned by disbanding the IMF, World Bank and WTO and passing the power to UNCTAD is fanciful.

A better approach would be to persuade the most powerful countries to support and lead reform. First, they need to be persuaded of the need for reform. Then the process must proceed to dialogue in which nations feel engaged and that their concerns are listened to, so that the structure that emerges has credibility and wide support. The overall concept is to reduce global control and encourage the emergence of regional and national solutions. The process of developing the new structure of international institutions should reflect the new way of thinking. Leadership could come from any quarter, but as the most powerful countries will continue to provide the bulk of the resources, they will have a strong influence. Where leadership is lacking, or agreement proves to be elusive, then the best course of action would be to pursue regional initiatives. For example, countries that feel marginalized should make efforts to pay off any outstanding loans to the IMF and World Bank and set up regional cooperative arrangements along the lines discussed in Chapter 9.

Bringing the Environment Centre Stage

The UNEP is the world's leading agency for the environment. Its mission is 'To provide leadership and encourage partnership in caring for the environment by inspiring, informing, and enabling nations and peoples to improve their quality of life without compromising that of future generations' (UNEP 2009). The influence of the UNEP has grown since its inception in 1972 and received a strong boost in 1992 from the UN Conference on Environment and Development, known as the Earth Summit. Despite regular reaffirmations of support for the work of the UNEP, the UN Secretary General reported to the Millennium Summit (UN 2000b): 'We must face up to an inescapable reality: the challenges of sustainable development simply overwhelm the adequacy of our responses. With some honourable exceptions, our responses are too few, too little and too late.'

In the first decade of the new millennium, there has been progress in raising awareness of climate change, but environmental issues are still not at the heart of world decision making.

The world does have access to expertise in how to shape society so that it has less impact on the environment. One such centre of expertise is the UNEP's Division of Technology, Industry and Economics (DTIE).

> *The DTIE works with international and non-governmental organizations, national and local government, business and industry to develop and implement policies, strategies and practices that are cleaner and safer, incorporate environmental costs, use natural resources more efficiently, reduce pollution and risks for humans and the environment, and enable the implementation of conventions and international agreements (UNEP 2009).*

The DTIE is small in size and influence. These kernels of expertise should be nurtured so that they can germinate into fully fledged organizations with the influence and credibility to facilitate the transition to a sustainable society.

The challenge is to translate four decades of words about the need to protect the environment into policy that will be effective in the real world. Expertise has to migrate from obscure offices inside the UNEP to positions of real power and influence. There are a number of ways to do this. The prime requirement is that world leaders agree to support changes aimed at delivering a sustainable world society – without preconditions. The world's experts can then come together to work on the challenge without their hands being tied by ideological constraints and without being influenced by vested interests. To indicate the direction that global cooperation should be taking for the environment to take centre stage, I present some ideas for consideration.

Reform of the IMF

Chapter 9 considers finance and capital flows and recommends greater national control of finance and economies. This would reduce the role of the IMF as countries followed this advice and took greater control and accepted greater responsibility. The transition would be smoother if the IMF were to lead the process by adopting policy that was less intrusive in national affairs but more focused on eliminating global imbalances. In a stable world financial system, a country's foreign exchange account should balance over the long term, subject to short-term adjustments to maintain equilibrium. When large imbalances build up, such as the huge debts of the United States or the huge surpluses of China, the IMF should intervene. The IMF should also be on the lookout for excessive dependencies in the system so that national economies are sufficiently isolated to maintain macro stability of the international financial system. In a system of independent economies, financial crisis can be isolated and a relatively small pool of IMF reserve funds should suffice. The IMF should also encourage regional arrangements to cross-hold currencies between neighbouring economies.

These changes do not conflict with the original purpose of the IMF and should be seen as bringing finance back under control. It is sensible that the world's strongest economies, with the most capability to inject additional resources, should have the greatest influence. However, the IMF should not be used to push ideology. World finance should be simple, stable and apolitical.

The WTO – Reform or Replace?

The World Trade Organization (WTO), established in 1995, is a forum for governments to negotiate trade agreements and settle trade disputes. It replaced the General Agreement on Tariffs and Trade (GATT), which was formed in 1947. The WTO operates a complex system of trade rules that have evolved through eight rounds of world trade talks. The raison d'être of the WTO is to liberalize world trade and this principle underlies all of its agreements. This makes the WTO the world-level organization that is most out of synchronization with a sustainable policy framework.

The conflict between aspects of free-trade policy and sound environmental policy was recognized by the establishment of the Committee on Trade and Environment (CTE) under the Marrakech agreement of 1994. CTE's terms of reference include: 'To identify the relationship between trade measures and environmental measures, in order to promote sustainable development' (UNEP 2005). At CTE's inception, the focus of policy was on supporting development in poorer countries, and export-led growth was considered to be the best way of making progress. The time has come to focus on sustainability, not 'sustainable development'.

CTE's terms of reference also include: 'To make appropriate recommendations on whether any modification of the provisions of the multilateral trading system is required, compatible with the open, equitable and non-discriminatory nature of the system' (UNEP 2005). This seems to have been drafted to counter concerns that the benefits of free trade could be hijacked by the environmental lobby. The statement 'compatible with the open, equitable and non-discriminatory nature of the system' means, in effect, that the principles of free trade are not negotiable.

In fact, what has happened is that supporters of free trade have hijacked the environmental agenda. Their ruling logic is that free trade is good for economic development, and that sustainability enters policy as a parallel mechanism. Sustainability has to become the ruling logic. Then, trade is allowed to adjust to an exchange of commodities, goods and services within a sustainable framework.

The Doha Declaration of 2001 focused CTE's role on the relationship between the WTO and Multilateral Environmental Agreements (MEAs) with a particular task of 'reducing or eliminating barriers to trade in environmental goods and services'. CTE was also instructed to give attention to 'the effect of environmental measures on market access and the environmental benefits of removing trade distortions' (UNEP 2005). The Doha requirements were not an invitation to CTE to carry out an independent assessment of how to resolve the inherent conflict between the WTO and MEAs.

Trying to add additional rules to the existing WTO agreements is like trying to divert a supertanker to a new course by throwing a line from a sailing boat whilst the supertanker remains under strict orders to continue full-speed ahead.

There is a fundamental conflict between the policy of free trade and sustainable policy. Interestingly, the reverse policy of seeking to close down world trade is more likely to lead to secure sustainable societies, but such a policy would also be ill-advised. The true need is for a policy framework for sustainable world trade.

Negotiations in the Doha round of world trade talks, which started in 2001, have developed into a stand-off between the developed and the developing nations. The main sticking point, which brought about the collapse of talks in Geneva in 2008, was a dispute over the Special Safeguard Mechanism (SSM). This measure was designed to protect poor farmers by allowing countries to impose a special tariff on certain agricultural goods in the event of an import surge or price fall. In particular, India wanted to be able to protect its poor farmers and so, together with China, could not reach agreement with the United States.

The impasse in the Doha round of talks is a clear signal that the appetite for liberalizing trade has waned. There are fundamental ideological differences leading to irreconcilable negotiating positions. The analysis in this book leads to the conclusion that free trade is no longer an appropriate policy; many politicians know this through gut instinct.

Effort should be directed at developing a replacement policy framework. There is little point in trying to add further complexity to a set of agreements that is based on the principles of free trade. Fundamental change is needed. The work of crafting a new trade policy must be insulated from the ruling logic of the WTO. This could be achieved through a separate project under the WTO umbrella, or – if this did not provide sufficient impartiality or analytical freedom – a new organization. A skeleton for such a possible new organization is presented below.

World Sustainable Trade Organization (WSTO)

Green campaigners have called for managed sustainable trade. Colin Hines proposed a General Agreement on Sustainable Trade (2000) which was then taken up by Woodin and Lucas (2004). Their proposal incorporates fair trade rules covering: 'guaranteed decent wages, working conditions, environmental standards and fair prices for producers and consumers'. I do not repeat the detail of their proposal because I do not believe that it would garner widespread support. I suggest that the core purpose of the WSTO, and the first article in its terms of reference, would be to ensure that international trade takes place in circumstances where:

> *the importing country is confident that the commodity or goods have been extracted from a sustainable source or produced in a sustainable manner, backed up by credible expert advice, assurances and guarantees.*

Such a system of world trade would be very different to the regime now in place. Countries that embrace the principles of sustainability will both enforce sustainable policy within their borders and influence the adoption of sustainable practices in other countries with which they trade. This would be the bedrock of sustainable world trade. Countries not committed to sustainability, or suffering from weak governance or ineffective enforcement, would be unwilling or unable to comply. Where these are poor countries, they will need to be helped with advice and assistance. If the country operating the trade is culpable of deliberately flouting the rules then, where aid is being supplied, the threat of the withdrawal of aid can be used. Countries that choose to stay outside the scheme may be able to exploit cheaper supplies in the short term, but being shut out of the most affluent markets is likely to be enough of a disadvantage to bring them into line.

The actions of the WSTO would be likely to lead to reductions in world trade as countries turned to their own resources as the first and default choice. On the other hand, new country-to-country trade channels would be set up as countries planned long-term reliance on resources not available within their own borders. Many economists will argue that the world economy would suffer a severe downturn, but I suggest that, as policy makers and businesses learn the new parameters, national economies will bounce back. The economic parameters may have changed, but business, in particular, is very good at exploiting new circumstances.

Under the regulations of the WSTO, the developed nations would be able to implement sustainable policy without suffering economic disadvantage, whereas the current world trade system discourages discrimination on the grounds of methods of extraction or production. Under the new system, countries that implemented strong sustainable procurement policies would have a commercial advantage over less scrupulous countries.

The development agenda for the poorer countries would shift from a focus on export-led growth without concern for the environmental consequences to building sustainable economies. This would support the true nature of sustainable development by leading to improvements in society rather than increases in industrial output.

Reforming Aid

As the world retreats from the policy of supporting development in poor countries through increased industrialization and export-led growth, a new aid paradigm is required. This is a huge subject that cannot be fully addressed here. Instead, I outline some of the areas where a change of emphasis is required.

The key change required in the delivery of aid is to retreat from interventionist and prescriptive policy. Outsiders arriving with a 'we-know-best' attitude may have good intentions, but they will fail to build sustainable societies that are in control of their own affairs, in tune with their local resources and proud of their communities. This is particularly true of Western aid to Africa, where it may be that less aid can deliver better outcomes (Glennie 2008).

Instead of delivering aid based on advice and intervention from countries with a very different heritage and culture, a better method would be to encourage cooperation within regional networks of countries. A country with problems could be teamed with a relatively more successful country with a compatible culture, similar resources and geography, and which faced similar challenges. For example, relatively successful South Africa could be teamed with struggling Zimbabwe. Aid from countries outside the region could go into supporting the partnership. This can be likened to advice from an understanding friend, backed up by resources from a rich distant uncle.

All aid delivered as money is vulnerable to corruption. One way to avoid this is by giving aid in other forms. When I worked in Africa in the 1980s, aid money was flowing in to help deliver clean water to villages. I observed how a commercial company maximized its profits by importing Western technology, paying appropriate 'fees' to middlemen, with a result that was of little benefit to the villagers. I was working as a young engineer at the time and could see much better ways of delivering appropriate solutions. Although not directly involved, I remonstrated with the company officials, but they were adamant that their responsibility was to maximize the value of the contract to the company.

I would like to see some grants of aid money diverted to directly employ qualified professionals at a similar salary to that offered by a commercial company. These professionals – who would typically be working in the areas of sustainable technology – would then be able to live within communities and share their expertise without strings attached, leaving the community in charge. This would have the side-benefit of exposing Western professionals to other cultures, giving them valuable insights that will influence their decisions as they progress in their careers. Such aid would also support the proposals in Chapter 10 to increase knowledge aid.

Some types of aid that are in favour now – for example, aid designed to increase exports or increase inward investment – will be employed far less. Such aid perpetuates the old idea that the focus of development should be on growing an industrial base to serve an export market. An increase in sustainable exports might be one of the beneficial results of providing appropriate aid and expertise, whereas a single policy focus on exports can all too easily lead to perverse outcomes, for example supporting the development of mining operations in a country with weak environmental regulations or weak enforcement. Another example is building factories for dirty processes that have become prohibitively expensive because of tight environmental regulations in the developed world. Such outcomes will be prevented when aid is kept firmly within a sustainable policy framework.

Loans to poor countries will also be less common as the problems of chronic debt are better understood and action is taken to protect these countries from irresponsible lending (as explained in Chapter 9). The lending role of the World Bank will be dramatically reduced and the organization will have to review its future role as the loan book is run down.

Conclusion

I have put forward an ambitious agenda to improve global cooperation and coordination. It is ambitious in the sense that I propose policy which opposes the tenets that have underpinned economic policy for three decades. My ideas will generate resistance and opposition. I accept that, from the narrow perspective of pure economics, I am championing turning back from further expansion in output, consumption and success as measured by GDP. This necessary retreat will lead the way to a better world society through the process of reformulating policy around sustainability. Instead of facing the seemingly insoluble problems of climate change, population growth and the end of the era of oil, there will be a realistic prospect that humanity will find solutions. World policy will then support efforts to build sustainable societies and will provide the context within which governments can tackle the complex and difficult issues of the twenty-first century.

12

The Key Role of Government

National governments are key to enabling world society to become sustainable. It is at this level of authority that real power resides. Governments have the hard power of executive authority, coupled with the soft power that a nation state possesses to draw on its population's commitment and loyalty to accept compromises for the common good.

A sustainable society needs intelligent government structured so that it can make difficult decisions that may be unpopular over the short term. More logical analysis and less political spin are required. The structure of the government should have an appropriate balance of power between appointed officials and politicians. There have been a number of encouraging developments. In the UK, the Sustainable Development Commission (SDC) has been given more independence (SDC 2009a).[1] In the United States, President Obama has appointed climate experts to key positions in his administration (see p. 162). These should be the forerunners of change in the way government tackles sustainability.

National Priorities and Local Circumstances

Each government has a unique set of challenges encompassing all the complex issues that affect its society. This complex web has to be brought into balance to build a sustainable society. Balancing national needs with national resources is at the core of policy making, but this does not go deep enough. The thinking that underpins comprehensive sustainable policy must include the flows of resources across borders to ensure that the other side of the transaction – which is beyond direct executive control – fits into a sustainable framework.

---

1 The Sustainable Development Commission was established in 2000. On 1 February 2009, the Sustainable Development Commission (SDC) became an executive non-departmental body (Executive NDPB).

As discussed in Chapter 3, it is beneficial to world stability and sustainability that there is variety in the way that countries are managed. Countries set their own priorities to suit their local circumstances, resulting in different ideologies, methods, policies and systems. In deciding how to run its affairs, a government has a range of options to observe and consider. Systems with a good track record, and operated by countries with a similar set of circumstances, may be copied. Countries that fare less well may be a showcase for policies to avoid.

It would be very dangerous to attempt to create a world society of nations cloned from one successful model. Any tiny flaw – human societies will always contain flaws – would be amplified across all the societies on the planet. There may be certain specific systems or policies that are adopted widely because they are so evidently good policy, but the total mix of policies will be different in each country. Identical societies are not possible, of course, because each country faces unique circumstances requiring a unique balance. It is fortunate for the stability of world society that differences are inevitable.

As the concept of 'strength in variety' gains official credence, institutions such as the IMF will back off from prescribing the same policy package to all countries. The various economic systems and regimes that already exist will be supported, and where necessary helped, but this will be done in response to local need and taking into account the prevailing culture, thus reinforcing national identity and sense of community.

Cuba, which under Fidel Castro has become a showcase of alternative policies, is an example to illustrate my point. From a Western capitalist perspective, Cuba is an economic basket case. No one wishing to retain their place on the faculty of a Western business school would dare to champion Cuba as a model, but it should be possible to look at the country with an open mind. For example, Cuba features at sixth place in the Happy Planet Index (HPI), an index that measures ecological efficiency in supporting human well-being.[2] The UK is at 108th position, the United States is down at 150th and Zimbabwe is at the bottom in 178th place (Marks et al. 2006). This insight should encourage Western policy makers to hold back on the instinct to deride Cuba and pause

2 The Happy Planet Index (HPI) is an index of human well-being and environmental impact that was introduced by the New Economics Foundation (NEF) in July 2006. Each country's HPI value is a function of its average subjective life satisfaction, life expectancy at birth and ecological footprint per capita.

to consider whether there are aspects of Cuban society that Western societies could copy.

In making the arguments in this book, my intention is to be apolitical. The world is governed by a variety of regimes ranging from dictatorships to democratic governments leaning left or right of the political spectrum. Some countries favour 'big' government with the tentacles of the state reaching deep into people's lives; other countries favour the concept of 'small' government in which the control of the state is minimized. All regimes, of all political shades, will face the challenge of implementing sustainable policies.

The current generation of world leaders, whose experience during the formative years of their careers was one of expansion and plentiful supply, will find the concept of sustainable living hard to grasp. Sustainable thinking does not come naturally to people raised within modern Western society. Support is growing for green policies, but it is hard for busy people at the centre of power to retreat from policies that have led to the economic success of recent decades.

Future generations will learn about sustainable living at primary school. In the future, sustainable decision making will be based on an ingrained, almost subconscious, understanding of the correct choices. The idea that policy frameworks should be sustainable will be as clear and obvious as our current understanding that financial budgets have to balance.

Politicians and government officials of today face a two-pronged difficulty. First, there is the difficulty of identifying sustainable policy. Decisions cannot be isolated into single policy choices. A sustainable policy framework is a complex mesh of interconnected policies. It is necessary to have a vision of the required end-state to know how to adjust each of the policies. Second, the public have little experience of sustainable living and it is, therefore, hard to persuade them of the need to change, especially where the required changes place constraints on behaviour and alter lifestyles.

A sustainable society will be a much better place to live, but until people have experienced it, and understand what is involved, attention will be drawn to those areas where restrictions are required. Resistance will be orchestrated by those who will suffer the greatest disruption in their lives – who will also be the people who have done the least to reduce their impact on the environment.

In due course, peer pressure, regulation and fashion will all work in unison, but this will take time.

There are times when democracy struggles to cope. The politicians do not know where they should be leading us and the public do not know what they should support for a better life. The expression 'the blind leading the blind' has never been more apt. Democracy can become stuck in a loop of short-term self-interest, with politicians looking for policies to announce that will appeal to voters and the electorate making decisions based on immediate impacts. It is vital that this political process is improved in order to be able to provide leadership through the difficult decades ahead.

Looking For Early Wins

An example of an encouraging development is the UK Government's decision in 2008 to merge energy policy with climate-change mitigation policy to create the Department of Energy and Climate Change (DECC). It remains to be seen how quickly this new department will achieve substantive progress, but there should now be a strong focus on solving the dilemma of finding the energy to run society whilst reducing carbon emissions.

In looking for solutions, the DECC will concentrate initially on finding renewable sources of energy to satisfy demand. It will soon become clear that sufficient renewable capacity will be hard to come by. Attention will then shift to influencing policy in other government departments. The solution to the dilemma of energy consumption and supply is not to be found in either energy policy or increasing the costs of carbon emissions. These are necessary but not sufficient. The solution is to be found in coordinated policy that reaches across all departments and which changes the whole infrastructure of society to become less energy intensive. City design, community design, transportation policy, building regulations – all must change.

In theory, there are simple mechanisms available, but often these will meet with resistance from the electorate. For example, in order to reduce CO_2 emissions, taxes on fossil fuels could be increased to such high levels that people are forced to react. This is a simple and neat solution. Necessity and hardship would encourage innovation and support business planning to enable a very different infrastructure to emerge. The rate of taxation on fossil fuels would be escalated over a timescale that follows the advice of climate scientists until fossil fuels are ultimately unaffordable. This simple mechanism, if taken in isolation, would also produce a backlash from the electorate. Governments have to think beyond such simplicity and identify the key elements of infrastructure change required for a low-carbon society in advance of increasing taxation on fossil fuels to unaffordable levels. An ideal solution is to announce the tax escalator at the same time as publishing the infrastructure plan, with tax revenue earmarked to fund the investment. This requires coordinated thinking and planning on a scale and of a complexity only normally seen at times of war.

Sustainable Policy in the Media Age

The government of a country at war tends to enjoy widespread support as the population rallies behind efforts to deal with national crisis. Criticism is often muted and seen as unpatriotic. For a peace-time democratic government, the situation is very different. The media age places unremitting pressure on politicians to justify their every move under the watchful eye of a critical media with 24-hour schedules to fill. This media goldfish bowl is a tough place to develop sustainable policy.

In the age of 24-hour television and the Internet, successful democratic political parties have become very adept at the processes of winning elections and staying in power. Media-savvy spin doctors work out how to target particular groups and advise politicians what to say. Such processes are not confined to elections but continue throughout the term in office. Opinion polls are monitored and announcements timed carefully to maximize the impact of good news and bury information that might reflect badly on the government. The input from focus groups seems to matter more than logical analysis. The art of political spin seems to have replaced firmly held principles and beliefs.

The mechanisms of modern government in the media age are not conducive to developing sustainable policy. Sustainable policy requires long-term planning and the implementation of policies that may be unpopular over the short term. The process of fixed terms of government provides a window of opportunity for bringing in unpopular measures, but when the next election looms, political spin once again takes over. A way must be found to bring long-term sustainable policy making to the heart of government.

Bringing Experts into Government

It would be presumptuous to believe I could design a better political system, but it would be appropriate to offer some advice regarding the future shape of government in a sustainable society. I believe that, over the coming decades, a greater role must be given to experts and appointed officials to ensure that logical analysis and long-term planning are utilized to build a sustainable society.

The appointments made by Barack Obama as he started his term in office are an example of the way that the sustainable policy debate can be invigorated. He selected climate-change expert John Holdren to be his chief scientific advisor,[3] respected climatologist Jane Lubchenco to head the National Oceanic and Atmospheric Administration (NOAA), and Steven Chu, an advocate for alternative energy, to be the Secretary of Energy. With such experts inside the US Administration, President Obama has laid the foundations for real progress in the important areas of climate change and energy.

It is very hard for government officials who built up their understanding of policy over the last two or three decades to grasp the enormity of the changes required and be able to propose alternative policies. Senior busy officials do not have time for the deep and radical thinking required. They are too close to the issues and too distracted by the pressures of running the current policy framework.

Increasingly, experts from outside government, who have developed a deep understanding of alternative policy, will be brought in to lead change. A very effective approach is to set up organizations where policy can be developed outside government, and which then influence and orchestrate change.

3 Harvard physicist John Holdren was confirmed as Assistant to the President for Science and Technology, Director of the Office of Science and Technology Policy, and Co-Chair of the President's Council of Advisers on Science and Technology on 20 March 2009.

The Sustainable Development Commission (SDC)

The SDC was set up by the UK Government in 2000 in response to the need for concerted action to progress sustainable development (SDC 2009a).[4] In 2005, the UK Government's Sustainable Development Strategy expanded the role of the SDC from advice and advocacy to being the 'watchdog for sustainable development'. On 1 February 2009, the SDC became an executive non-departmental public body (NDPB), further reinforcing its remit as the UK Government's sustainable development watchdog and advisor.[5]

The SDC agreed with the UK Government a common approach to assessing whether a policy delivery is sustainable according to 'Five principles of sustainable development' (SDC 2009b):

- living within environmental limits;
- ensuring a strong, healthy and just society;
- achieving a sustainable economy;
- using sound science responsibly;
- promoting good governance.

The SDC's principles are likely to fit any national sustainability initiative, with modification if necessary to reflect the particular challenges that a country may face.

The SDC is well placed to develop sustainable policy, because it can draw on a wide range of experts and, as it is outside direct government control, has the independence to tackle sensitive and difficult issues. However, SDC does not have direct executive authority to impose sustainable policy.

An example of SDC's work is the 'Breakthroughs for the Twenty-First Century' project (see Box 12.1 overleaf).

4 The UK Sustainable Development Commission (SDC) was set up in October 2000. It was preceded by earlier initiatives: the UK Round Table on Sustainable Development and the British Government Panel on Sustainable Development.

5 The SDC was part of DEFRA until 1 February 2009 when it became an executive NDPB, giving the SDC increased independence from government.

BOX 12.1 BREAKTHROUGHS FOR THE TWENTY-FIRST CENTURY

The Sustainable Development Commission (SDC) launched a project in 2008 to invite people from any part of society to contribute breakthrough ideas. At the launch, the then chairman of the SDC, Jonathon Porritt, said:

'We at the Sustainable Development Commission want to bring together the most compelling and creative of these ideas – those that we think can really help us move forward. We want a dynamic and hard-hitting collection of breakthrough ideas brought together in one place that will really inspire and motivate policy makers and others to catalyse change.'

A breakthrough idea could be a new way of thinking or working, or a new technological solution. Previous suggestions that for some reason had not been taken forward were also encouraged, together with ideas that had already been applied but needed scaling up. The breakthrough could be how to get an idea to happen, rather than the idea itself.

Out of 285 ideas submitted, 19 were selected for the final report launched at an event in London on 1 July 2009. They covered three areas: sustainable lives, sustainable places and sustainable economy. The ideas ranged from city agriculture and making cycling mainstream to a range of low-carbon solutions and the idea of green bonds to mobilize investment. (SDC 2009c).

The 'Breakthroughs for the Twenty-First Century' project is an example of crafting an environment in which radical ideas can be considered and exposed to dialogue and discussion without getting killed off before the possibilities have been fully explored.

Transforming society to become sustainable requires an overarching commitment to find sustainable solutions to which all government departments conform. The SDC has proved to be a useful organization to influence policy and champion change. The SDC regularly issued independent reports that challenge the Government to take action (Ullah et al. 2009).

The key to success is a balance of independence from government (in order to be free to consult widely and not to be constrained when searching for solutions) and closeness to government in order to be able to influence policy. During the period of the Sustainable Revolution, when officials will struggle to understand sustainable policy, organizations such as the SDC will have a vital role. In 2010, the SDC is under threat of having its budget withdrawn. Organisations such as the SDC will have to work hard to demonstrate the delivery of improvements and savings in the near-term in order to survive.[6]

The SDC was set up to lead the UK towards improved sustainability credentials, but worldwide other countries are setting the pace. Policy makers should turn to the countries at the top of the Environmental Performance Index (EPI)[7] for ideas and policies on how to incorporate sustainable policy. In 2008, Switzerland, Norway and Sweden were top of the EPI (the UK was in 14th place and the United States was in 39th place out of 149 countries) (Esty et al. 2008). Other countries that are consistently high in the EPI are Finland and New Zealand. It is beyond the scope of this chapter to provide a country-by-country review of methods and progress. Each country is developing sustainable policy in its own way.

Another organization that has been successful in developing sustainable policies from which lessons can be learnt is the European Commission.

Lessons from the European Commission

In 1997, sustainable development became a fundamental objective of the EU when it was included in the Treaty of Amsterdam as an overarching objective of EU policies. At the Gothenburg Summit in June 2001, EU leaders launched the first EU sustainable development strategy. This 2001 strategy called for a new approach to policy making that ensures that the EU's economic, social and environmental policies are mutually reinforcing. The strategy was based on a proposal from the European Commission.

The term 'European Commission' refers to the 27 commissioners appointed by the EU member states after approval by the European Parliament. The Commission is supported by approximately 38,000 staff. The role of the Commission is to draft new laws and regulations, which are then submitted to the European Parliament and the Council of Europe for debate and decision. The Commission is also in charge of the day-to-day management of EU policies and activities.

6 On 22 July 2010, the UK Government announced that funding was to be withdrawn from the SDC from 2011.

7 The Environmental Performance Index (EPI) has been published for 2006 and 2008. It was preceded by the Environmental Sustainability Index (ESI), published between 1999 and 2005. Both indexes were developed by Yale University (Yale Center for Environmental Law and Policy) and Columbia University (Center for International Earth Science Information Network) in collaboration with the World Economic Forum and the Joint Research Centre of the European Commission.

The European Commission is staffed entirely by appointed officials and, although formal power resides with the Council and Parliament, the Commission wields considerable influence. There are some complaints that the Commission is too powerful and is not sufficiently accountable to either the European Parliament or the member's national governments. In reality, this is one reason why Europe leads the world in developing and adopting sustainable policy.

The Commission is one step removed from the political process. This allows policy to be developed and drafted on the basis of logical analysis at a distance from political considerations. The Commission has drafted legislation for a number of sustainable policies, and then championed their adoption, such as the Waste Electrical and Electronic Equipment Directive (WEEE Directive) which came into force in January 2007. These are not vote-winning measures; elected politicians may have had difficulty finding the time to give them priority. Once legislation reaches draft form, it is discussed, amended and approved by the elected parliament, but the fundamental thinking has already taken place inside the Commission.

The Renewed EU Sustainable Development Strategy was adopted by the European Council in 2006. The stated key objectives are (Council of the European Union 2006):

- Environmental Protection
- Social Equity and Cohesion
- Economic Prosperity
- Meeting Our International Responsibilities

These objectives could be applied to any country with the underlying detail designed to reflect local priorities and circumstances. The order in which they are listed illustrates the priorities that emerge when the focus shifts to sustainable policy. The priority becomes protection of the environment and maintenance of social cohesion followed by economic performance.

The EU Sustainable Development Strategy is not the only policy in Europe but it is vital that it is used as an overarching policy. For example, the implicit assumption in setting economic policy is often that economic performance and economic growth can be used as proxies for improvement in society.

In a resource-constrained world, this assumption no longer applies. The first objective of the EU Sustainable Development Strategy, environmental protection, includes the necessity to 'break the link between economic growth and environmental degradation'. This is important, of course, but the thought process has to go further with an acceptance that economic growth can no longer be the prime focus of policy.

Ingrain Sustainable Thinking in Government

A government's established economic advisors will find the shift of focus from economic growth to a sustainable policy framework hard to understand. Over time, the economic establishment will accept that their discipline has changed, but for now such ideas should not be pushed too hard. Low-growth or steady-state economic policy is feasible in a democratic-capitalist system (Lawn 2005), but forward-looking economists still need time to refine their models and win acceptance for them. If a government gives primacy to the sustainable policy framework, the pressure is on the economists to make their models suitable. This is the correct hierarchy of policy making, but it is different, and for that reason alone will generate resistance.

Sustainable policy formulation requires different thinking and action is now urgent if world society is to become sustainable before severe and irreversible environmental degradation is caused. The drafting of sustainable policy needs experts working at a distance from day-to-day government. Countries without a heritage of sustainable thinking may need to establish a sustainability commission which could draw on a wide range of experts and which had the power to influence policy. This will work only if departmental officials are willing and able to drive the changes required.

Moving sustainable thinking forward from advocacy to action will require the appointment of particular experts to senior roles within government departments. Their remit will be to connect sustainable policy. Such appointees would need to be credible experts, perhaps outspoken critics. Some of the changes required in the way society operates are revolutionary and will require the disruption of long-established processes. People who can speak out and shake up the system will be useful until the system settles into a new and sustainable policy framework.

Politicians will, of course, retain a veto over decisions made by officials. This is necessary to maintain the principle of democratic government, but politicians have to be persuaded to appoint experts to powerful positions and then allow them considerable scope to lead change. The main role of politicians is to give unequivocal high-level support to the principle of sustainable living and to defend that principle to the electorate. The appointed officials can then take the tough decisions required and draw criticism for their actions away from the elected representatives.

If politicians are drawn in close to the issues of implementing sustainable society, and have to shoulder complaints about each stage in the process, political motives will intrude and progress will be slowed. I suggest that there should be fewer government ministers and more appointed officials.

There is a danger in this model. Unelected organizations like the UK's SDC will meet resistance as they move beyond the obvious actions and easy wins to tackle difficult and politically sensitive issues. Examples of such issues are enforcing full life-cycle production and consumption of products, and stiff restrictions on the ownership and use of cars in urban areas. It could be hard to implement such measures without a democratic mandate. Politicians will have to be persuaded not to veto unpopular policies, and they should also encourage officials to search for policies that mitigate people's concerns, such as improvements in public transport.

In addition to resistance from the electorate, there is likely to be resistance from established experts. Economists, business school academics and experts in public administration will have to be persuaded to develop and use different models. There are sacred cows within business schools and mainstream economics that will not survive close examination as policy makers truly understand the implications of adopting sustainability. It will require deft handling to maintain the pace of progress without alienating old-school experts, who will have to adjust to a new reality.

When it becomes clear that governments are serious about adopting sustainable policy, I believe that the business community will be the area that offers the least resistance. Business leaders will have to be persuaded, along with everyone else, but they tend to be open-minded when looking for opportunities during periods of disruption. I believe that business can respond quickly and become the primary agent for change if governments take control of policy in the way outlined in this chapter.

13

Global Corporations

Enlightened self-interest is the essence of corporate responsibility and the key to a better world.

Ban Ki-moon (2009)

In the era of globalization, multinational corporations (MNCs) are accused of becoming a law unto themselves, disconnected from government control and accountable to no one. An MNC will stay in a country whilst its operations are profitable, but will depart when profits dip or better investment opportunities present themselves elsewhere. There is little point in complaining; this is the nature of economic globalization and MNCs are simply behaving exactly as the theory of open markets suggests they should.

Within a sustainable world, more will be expected of MNCs. They will need to learn new ways of operating to shake off the label of 'big bad business' (unfairly but often applied) and become 'big intelligent business'. Forward-looking business leaders who have been following developments by reading works such as *Factor Four* (Weizäcker et al. 1998), *Natural Capitalism* (Hawken et al. 1999) and *The Natural Advantage of Nations* (Hargroves and Smith 2005) will be well informed about the radical thinking required. Chief executives who see no further than the current Corporate Social Responsibility (CSR) agenda will struggle. Marc Epstein (2008) describes very well the processes that CSR professionals have developed to 'make sustainability work'. This approach has had some success engaging with business, but few business leaders truly understand the opportunities that will arise as the Sustainable Revolution takes hold. Those who do will be able to grow sustainable long-term profitability.

There will be businesses that take a cynical view of exploiting change without adopting the principle of sustainability, for example planning to profit from the emerging carbon market without buying into the underlying purpose of reducing reliance on fossil fuel. Now, there are profitable opportunities to exploit an imperfect system. Over time, government will find ways to close the loopholes and the companies simply playing the market for profit will be forced out. The businesses that understand the long-term need to reduce fossil carbon release – and have the methods, processes and technology to deliver reductions – will thrive.

There are also businesses that, by their nature, will not survive in a sustainable society. I categorize these as 'pariahs' (McManners 2008). In the unsustainable world of today, these companies are banking profits and the market gives the shares a value based on the expectation of future profits. The point will come when they are exposed as businesses that have no future. The change of sentiment could be rapid. The shares might be trading on an average multiple of last year's earnings one day and be worthless the next. There are companies with apparent value now that appear worthless when viewed through the lens of sustainable policy. An example is airlines that own large fleets of conventional aircraft and have no plans to migrate to the sustainable airline industry of the future. As more people are converted to the concept of sustainability, fewer investors will be willing to hold such shares. It will be hard to call the tipping point. Clever investors will divest ownership whilst the share is still able to generate a profit and therefore appears to have value.

The airline and aircraft industries provide an interesting insight into the nature of the current stalemate. Current operations are clearly unsustainable, but people like to fly and like flying to be cheap. It is as if there is a Faustian pact, in which we are all complicit, to continue to enjoy cheap flights knowing that the damage may destroy the beautiful locations we jet off to visit. The reason I find this so interesting is that it will be possible to build an industry around sustainable flying, but it will be very different to the industry we have now. The airlines and society are stuck in a stalemate, preventing the actions I describe in *Victim of Success* (McManners 2009: 124–8). When the current stalemate is broken, it is highly unlikely that the existing airlines will survive, because for far too long we held them to delivering cheap flights, and judged the management on delivering short-term profit targets.

The managers of pariah companies may seek to leave and find a job in companies that have a future in sustainable society. However, there is a job to be done to manage these companies in their twilight years, which requires good managers who understand the special position they are in. Oil companies, for example, will go out of business in due course (unless they reinvent themselves as renewable energy companies). There is likely to be another profitable decade in the oil industry before the approaching decline becomes serious. The CEOs have an interesting and challenging task to manage their companies, watching carefully for what world society will allow. I listened to a lecture by James Smith, Chairman of Shell UK. He spoke about the exploitation of oil sands and whether society will allow it. He has a commercial responsibility, if permission is given, to exploit these fossil-fuel reserves. Shell will do it efficiently and with due regard to local environmental impact. It is far better that a responsible and well-run company such as Shell carries this out than a less reputable operator. Of course, as a society we should reach agreement that such low-grade sources of fossil fuel are not to be exploited. But it is the job of all of us to make our views known, so that politicians act on our wishes. Shell and the other oil companies will do what *we* want.

At the time of writing, those of us campaigning for revolutionary change are a small group. If our views start to dominate the agenda, CEOs will make some rapid strategic shifts in direction. They will not act now because analysts and fund managers – who represent the bulk of investment funds – would not provide their support. There has to be a genuine shift in public opinion to spur change. Then business too will change.

Those who oppose big business and object on principle to their power and influence should reflect that business reflects society and conforms to public policy. If attitudes can be changed, and policy altered, business can be mobilized as the primary agent for change.

Moving Beyond Globalization

The policy of globalization and the rise of the MNC are inextricably linked as markets become increasingly global, as described by Friedman in *The World is Flat* (2005). Globalization would not have proceeded as far, or as fast, without the capability of MNCs to connect markets and build global supply chains. On the other side of the coin, MNCs would not have been able to gain such power and influence without the policies of free trade, deregulation and free capital flows. Globalization and the rise of the MNC are aspects of one narrative.

As the proximization policy framework reconciles the conflicts between a narrow focus on economic performance and sound environmental stewardship, globalization will change. The MNC will have to change too. The interdependencies and linkages are too close for MNCs to ignore the coming changes. As world society changes dramatically, MNCs will have to reinvent themselves.

MNCs have taken a proactive role in driving globalization over recent decades. In the next decade, business leaders will face a choice of defending the policies of globalization or driving the new agenda for change. History shows that companies that resist change become stuck in a rut that leads eventually to failure. Successful businesses embrace change and put themselves in the driver's seat. This gives them the ability to influence change to their advantage and also ensures that change happens more quickly than would otherwise be the case. When the new policy is sustainability, then what is good for the MNC is also good for society.

The Influence of MNCs on the World Economy

MNCs are powerful players within the world economy, influencing policy and generating economic activity. They can be a force to improve society if the framework within which they operate allows it.

MNCs attract accusations of bad behaviour because so much of what happens in the world is delivered by corporations. When it goes wrong, it is natural and logical to blame them. However, the better target for complaint is often the government policy framework (or lack of framework) in which they operate. If the framework is improved, the MNCs can be mobilized to transform society and the world economy.

In the current policy framework, governments look favourably on MNCs when they bring inward investment to build factories and provide jobs. There may be grants, subsidies and soft loans to attract such investment. The quantity of foreign direct investment (FDI) is used as a measure of how much support and stimulus is coming into a country to help it develop and grow its economy, but this is too narrow a measure. In a sustainable world economy, FDI is only effective if backed up by sound analysis of the long-term impact.

There are examples of international corporations entering a country bringing investment, development and jobs, only to leave again when the opportunity has been milked dry. This situation is particularly worrying when the country is poor with a lack of effective governance, and the attraction for the MNC is non-renewable natural resources such as copper or oil. A mining company can move in, build a mine, exploit the deposit and withdraw when the mine is depleted. To protect the corporation, a local company will be set up to operate the mine. In the early years, it will receive investment from the parent corporation. In later years, the corporation receives the cash flow from the sales of the ore. Little of the cash may pass through the country where the mine is located, unless the country acts robustly to claim a share through taxation or other measures.

When the MNC makes the decision to withdraw, it will starve the local company of new investment, milk it for cash and let operations run down. As the mine approaches the end of its useful life, the simplest option may be to hand over ownership of the local company to the government. The government would then be able to run the mine for a while longer, but the main attraction for the corporation is to divest the residual liability to clean up the site. The same outcome could result if a country starts to doubt the long-term commitment of the corporation and nationalizes the assets. If these are mature facilities that have already generated good cash flow over many years, nationalization may suit the corporation to release it from long-term liabilities for facilities built before high standards of environmental safeguards were implemented. As countries take closer control of their resources, these types of decision will be common.

It would be wrong and unfair to concentrate on the negative influence of corporations. Reputable MNCs can no longer afford to take the risk of exploiting low standards and weak enforcement in poor countries. The media's capability to find the information and expose the story is too great. The leading global corporations look beyond the commercial bottom line of a particular project to protect their long-term reputation. They do not want to lose sales through stories in the media of exploiting poor countries. They also want to get permission for new projects in other locations. It is in their commercial self-interest to run factories and facilities with standards that exceed those of the local companies. Such corporations can also push up the general standards of safety, welfare and environmental protection as they impose requirements on their local suppliers and contractors.

The intentions of the MNC as it interacts with the world economy are not driven by altruism, and their activities are not a form of aid. The MNC enters a country to turn a profit. In doing this, the MNC is more sophisticated than local companies, is more powerful and has more capability. This is why it is so important that the framework in which they operate encourages the positive contribution that MNCs can make and penalizes behaviour that causes environmental or social harm.

The Transition from MNC to Global Corporation

The first MNC was the Dutch East India Company founded in 1602. It was a highly profitable company during its two centuries of existence, not just for its own shareholders but also for the Netherlands, bringing great wealth to the country as it became a conduit for spice into Europe. It was the most powerful corporation the world has ever seen, with many powers normally reserved for government, including the ability to wage war, negotiate treaties and coin money (Ames 2008). The collapse, when it came in 1798 with huge debts, has uncanny parallels with the collapse of Enron in 2001 (see box).

BOX 13.1 THE DUTCH EAST INDIA COMPANY

The multinational corporation (MNC) is not a recent invention. It has been around ever since 1602 when the Dutch East India Company was established. The basis of its early success was a 21-year monopoly, granted by the States-General of the Netherlands, to carry out colonial activities in the area of Asia now known as Indonesia.

The Vereenigde Oost-indische Compagnie (VOC), as the company was called in Dutch, was a successful business for almost two centuries, paying an average annual dividend of 18 per cent. The Dutch East India Company was finally brought down when the first speculative market bubble in history imploded. The basis of the speculation was, surprisingly, tulip bulbs, the apparent value of which exploded under the influence of trading akin to modern options trading. This was the trigger that exposed other failings. Under pressure to make the numbers, the board had resorted to all kinds of financial engineering to keep up its notional shareholder returns. Cash dividends were increasingly replaced by bonds. In its declining years in the late eighteenth century, the company was referred to as Vergaan Onder Corruptie (referring to the acronym VOC) which translates as 'Perished By Corruption' (Ricklefs 1991). In 1798, the company collapsed with huge debts.

The ownership of the Dutch East India Company was predominantly Dutch with a small number of shares held by German immigrants. It also had a close relationship with the Dutch state. When the company collapsed, its assets became the property of the Dutch government.

There are many modern MNCs with strong national roots, but MNCs are increasingly global in both ownership and operations. Typically, an MNC is a public listed corporation (plc) traded on more than one stock exchange. The shares can be bought and sold by a wide range of investors from many countries. This ownership structure puts a strong responsibility on the executives of the corporation to manage it primarily to increase shareholder value.

Executives running any business also have to take into account the interests of a range of stakeholders. If the business is a national business, these stakeholders range from customers and suppliers to government and special interest groups. National companies have a range of loyalties to consider for long-term performance and they are wary of upsetting stakeholders: this preserves profitability over the long term.

Stakeholders other than shareholders have little real influence on an MNC operating in an open global market place, where there is considerable freedom to shift market focus or relocate production. These include national governments, who can find their interests sidelined in favour of maximizing profits at the global corporate level. Business schools reinforce this behaviour by teaching that shareholder value is the prime metric for measuring the success of corporate strategy. The business world of the twenty-first century needs more enlightened thinking than this.

The MNC has evolved from national companies with global outreach to become true global corporations. They may not have the same raw power as the Dutch East India Company, but they have enormous economic power and influence.

The Changing Business Environment

Global corporations may have a physical location for their headquarters but they do not answer to any one government. They are free to select which country to make their prime location, based on tax rates and regulatory requirements. Whilst the world operates an open globalized economy, governments have little real control over the biggest global corporations.

The balance of power is changing. Across a broad range of economic sectors from energy and power generation to telecoms and minerals, the state is taking more control. China and Russia are leading the way in the strategic deployment of state-owned enterprises, and a number of emerging-market governments such as Venezuela and Bolivia have begun to follow their lead, moving beyond regulation to direct intervention.

The world oil markets provide a good example of the shift to greater national control. Multinational oil companies now produce just 10 per cent of the world's oil and gas and hold about 3 per cent of its reserves. State-controlled companies are in charge of more than 75 per cent of these global crude oil reserves and the largest oil companies are all controlled by governments, including Saudi Aramco, Gazprom (Russia), China National Petroleum Corporation, National Iranian Oil Company, Petróleos de Venezuela, Petróleo Brasileiro and Petronas (Malaysia). The world's largest multinational oil company is Exxon Mobil, but it ranks at a lowly 14th in the world (measured by the reserves under management) (Bremmer 2009).

In the oil industry, multinationals continue to hold competitive advantages in development and production of deep sea and other technically difficult projects. This points towards the future for the global corporation, where the prime focus will be on knowledge management and innovation rather than mobilizing the brute force of advantages of scale.

The sustainable world in the twenty-first century will be different. The physical production, consumption and recycling of products becomes a local business. Comparative advantage is no longer to be found in the production of physical goods but in the knowledge, design and know-how that is required. The importance of the global knowledge economy will increase substantially.

The Twenty-First Century Global Corporation

The leading corporations will be the architects of their own transformation as they see the attitudes of government and society shifting. They will remain hugely influential with enormous soft power, but will also have to tread carefully in their relationships with sovereign governments or take the risk that government uses hard power to force change. There will be a complex interaction of issues impinging on the evolution of the new global corporations as efforts are made to put society on a sustainable track. It is impossible to predict the outcome with accuracy, but I put forward here a vision of how successful twenty-first century corporations will operate.

The corporate structure will provide considerable autonomy to national business units or subsidiary companies. This will reflect the structure of a sustainable society where the focus is on local balance in the application of resources and setting policy.

The proportion of local ownership of subsidiary companies will increase. Corporations that resist this transfer of ownership will risk having their operations nationalized, as President Evo Morales did in asserting control over Bolivian energy resources in 2006. An example of the attitude that corporations should expect from government is the Belgian Government's decision to retain a controlling interest in its leading telecoms company, Belgacom. Government will be content to accept the expertise of global corporations, and their injections of capital, but increased government control will prevent the corporation from making an early exit.

In the past, countries have competed for FDI in what can be characterized as a race to the bottom by lowering taxes and removing regulations. It will take time to reverse this trend into a race to the top. Corporations will continue to look for favourable treatment and, at first, increased regulation will be unattractive until the benefits start to flow. Over time, countries leading in implementing sustainable policy will become attractive destinations for corporate investment as these will be the premium markets and a source of the methods, processes and technology required to run a sustainable society.

Twentieth-century corporations had already evolved beyond the colonial presence of the eighteenth and nineteenth centuries. In the twenty-first century, this evolution will shift further away from the model of exploiting resources with a narrow focus on profit towards global corporations that behave more like sponsors, mentors and knowledge managers. It would be good for world society if this were to be so. In the globalized world of today, my vision appears naive but it is based on sound logic. All it takes is to shift the core principle of policy from laissez-faire capitalism to sustainable policy, capitalist or not. Governments will then make policy choices that support corporations that make the required transition and penalize corporations that resist. For example, corporations will be required to demonstrate that their material inputs come from sustainable sources in order to be able to sell in the markets of countries that have embraced the principle of sustainability.

The competitive advantage of corporations will change as twentieth-century supply chains are dismantled to adopt sustainable production and consumption. Searching out cheap production locations from which to ship goods across the world to the most profitable markets will no longer be a substantial source of profits. There will still be global supply chains for certain resources, such as energy, but they will be under close oversight as governments adopt sustainable policies.

As global corporations redefine themselves, they will look for new sources of competitive advantage. The most profitable opportunities will arise from leading change at a pace faster than society demands, developing technology one step ahead of each tightening of environmental regulation. Renewable energy technologies, for example, are receiving a flood of venture capital to support the development of affordable solar photovoltaic (PV) panels, third-generation biofuels and more efficient, quieter wind turbines. Capital is also pouring into the development of improved battery technology, because corporations see that the car of the future is likely to be electric, and that, at the moment, the Achilles heel of the electric car is the battery.

The most profitable businesses in a sustainable world have yet to be defined. If they were obvious, corporations would already be working towards exploiting them. The opportunities with the most potential are likely to require complex change and advanced technology not yet invented. This is where the scale of global corporations may be able to make progress towards solving some of the world's most difficult technical challenges – such as designing and developing the infrastructure to produce 'liquid sunshine'[1] in large quantities from the huge empty desert regions of the planet.

Global corporations will support, and even lead, the transition to a sustainable society because it will be profitable. In *Adapt and Thrive* (McManners 2008) I describe a method of developing green corporate strategy. My intention was to make a persuasive case that corporations should act early to support my conviction that world society must change. At the time I was writing, I was out of step with many business leaders and management theorists. Now, more and more corporations are adopting such an approach, not necessarily because they have read my book but because they have carried out the same logical analysis and reached the same conclusion. Society is on the verge of great change, and business must change too.

1 'Liquid sunshine' is the term I used for a renewable fuel from the desert without being prescriptive about what it might be. It is likely to be a biofuel, but it could also be liquid hydrogen or another high energy density liquid yet to be developed (McManners 2008: 67-8).

Many corporations have not yet understood the risk of failing to make a substantive move towards more sustainable operations. It is not simply the case that corporations have the option to exploit the opportunities of the Sustainable Revolution; there will be severe negative consequences for the corporations that lag behind. The example of General Motors (GM) illustrates this well. GM was once regarded as the rock of the US corporate world. The collapse of GM in 2009 was presented in the media as a consequence of the credit crisis. This hid the fact that GM sowed the seeds of its collapse when it decided to fight against action to reduce fossil-fuel consumption, instead of leading the United States away from its addiction to oil. From that point on, my strategic judgement was that failure was inevitable; only the timing was in doubt. Other car corporations, such as Toyota, have suffered in the recession, along with all companies, but Toyota's core business is sound as it continues to design and sell fuel-efficient vehicles based on hybrid technology. This is just the beginning: the pace of change in car design will accelerate. Any car corporation lagging behind now will find it hard to catch up.

Following the 2008 financial crisis, influential business advisors warned corporate managers that globalization may no longer be the dominant paradigm (Bremmer 2009). But underlying such advice was the assumption that when the recession ends, economic globalization will resume its course. Few people have taken the next step to accept that economic globalization is no longer appropriate policy for the world. Until business leaders, and the business schools that teach them, understand this major shift, the debate over the future of twenty-first-century corporations will not progress far. The debate will be stuck in a circular argument of attempting to recapture former wealth using the methods of the past and finding that these methods no longer work. Instead, the focus should be on building new wealth based on an evolving new paradigm, at the core of which is the policy of proximization I present in this book.

14

Human-Scale Communities

Capitalism is built on the assumption that selfish greed works as a mechanism for allocating resources and running society. However, the secret of human success is the ability to work together and display less selfishness behaviours in order to be able to share in common success. A stable and sustainable world would be dominated by human-scale communities in which 'selfish altruism' applies.

Human interaction is often irrational and illogical, as economists discover when markets fail because people do not follow logical economic choices. People can be angry at perceived unfairness and are prone to knee-jerk reactions to defend the status quo, despite sound arguments for change. People can also be generous and concerned for the welfare of others. It is fortunate that human nature does not follow the logic of conventional economics. It is not human nature that is at fault, but economics that is too narrow.

Sustainable living will transform the soulless megametropolises of the twentieth century into a patchwork of dynamic vibrant communities. Lives and relationships will be much more localized, enabling people to draw on the human ability to share, cooperate and reach consensus. Future communities will be more complex, include more variety and, I contend, be much more friendly and fun. This is not a weak or easy option: it will require strong leadership and robust action to face down opposition from people still stuck in a twentieth-century mindset.

The Transition Towns movement in the UK is an example of the coming changes as communities start taking action to prepare for the future when oil runs out. Rob Hopkins (2008) shows how the inevitable and profound changes ahead can have a positive outcome. The slow cities movement is another source of inspiration bringing together urban design and the social construction of place (Knox 2005). These changes can lead to the rebirth of local communities, which will grow more of their own food, generate their own power and build their own houses using local materials. They can also encourage the development of local currencies, to keep money in the local area (Chapter 9). Reshaping communities to fit a sustainable world society is to be welcomed, not feared.

The Power of Community

Each person contributes a small component to the complex structure of society. No matter how strong, capable or intelligent an individual may be, they will achieve nothing substantial unless their efforts are part of a coordinated effort with other people. The power of community affects every aspect of society from virtual communities of people with shared self-interest to the physical communities where people live.

Even activities that appear to be isolated, such as writing, are not. When I write, I sit alone with only a cup of coffee for company. As I corral my thoughts into a reasoned argument, I build up a vision in my mind of a better society and the policy changes required. These rough-cut ideas contain useful insights, but they have to be refined and communicated before they have real value to society. It is interaction with other people that develops the germ of an idea into a fully fledged concept. To become useful, the concept then has to evolve into policy. This is not a straightforward logical process, particularly when it relates to people's lifestyles and the design of the communities in which they live. Policy evolution has to take into account people's values, aspirations and, that most elusive of factors, emotions.

Successful human communities are a complex reflection of human nature. A factory can be designed to deliver products efficiently and effectively, but human communities are different. Human lives do not submit to logical sequential analysis.

Big impersonal bureaucracy, which is well designed and efficient, can be thwarted by human intransigence. From the bureaucrat's perspective, such resistance displays human failings and must be overcome. The alternative approach is to allow human instinct and behaviour full force to shape communities and set policy. Mistakes will be made – that is human nature – but the people involved have a direct self-interest in making adjustments and improvements. The power of this human process is that the required compromises are made within the community, and the solution is owned by the community. In the sustainable communities of the future, people will have to give up things that they had in the past. An example is the ownership and use of cars. In towns and cities in the developed world, the majority of residents own a car that is habitually parked outside their dwelling. Car ownership is so deeply ingrained in their psyche that it will be hard to implement policy that gives the car a lower status in city planning and design. If a whole street of car owners is told by a higher authority that they can no longer park in the street, this is likely to lead to coordinated resistance by the residents defending their 'rights'. An alternative approach is to initiate a dialogue about whether the residents would like their street to become a car-free zone, in which children can play in safety, with limited access for deliveries. This approach sets up a complex, ill-defined and unpredictable human dialogue. The result of the deliberations will have to be respected. Some streets may choose to maintain the status quo, but other street communities may become enthusiastic advocates for improving urban life.

Communities have the power to craft sustainable policy frameworks that balance social and environmental issues in a way that is economically viable. Crafting such policy requires the administrative capabilities of coordination and long-term planning, but it also requires human engagement to be able to compromise, cooperate and ration scarce resources. On occasions, building a sustainable community will require acceptance of hardship to make the transition to a better society. These are issues that are best dealt with by face-to-face negotiation between people who know each other and live within the same community.

Applying Community Engagement

There are numerous examples of how community engagement can lead to solutions. The problems may seem impossibly difficult until people can be engaged in dialogue, leading to the resolution of long-standing intractable conflicts, such as that in Northern Ireland. From outside the province of Northern Ireland, the conflict appeared to be a ridiculous waste of people's lives. No amount of well-intentioned logical analysis from people in London, Dublin or Washington could solve it. The people of Northern Ireland had to be brought together in a dialogue over their future. Tony Blair, Bertie Ahearn and Bill Clinton all deserve some credit for their involvement, but the linchpins in the solution were the local leaders such as Gerry Adams, David Trimble,

Martin McGuiness and Ian Paisley. The latter two were bitter sworn enemies on opposite sides of the dispute. The peace was cemented when they shook hands, made peace and entered into government together. Their respective supporters observed and followed. Human face-to-face engagement at the local level is the way to solve the problems of society. This applies to major conflicts and big issues as well as to everyday low-level decisions about managing society.

The disposal of rubbish in the shanty towns of South America is an example of a relatively minor problem that needs a solution. I cite this example because it clearly shows what not to do. I was part of a discussion in which a representative from the region presented his solution. He wanted to introduce a fleet of handcarts that would gather the rubbish from the narrow alleys and take it to a central collection point, where it could be collected by vehicle to go for processing. He explained that this would bring jobs to the community and had the advantage that the carts could be maintained and repaired using local tradesmen. From my time working in Africa, where I observed the informal settlements around Yaoundé, Cameroon, this seemed like a perfect solution. Of course, my opinion was not important. The importance of the proposal was that it was a local design, championed by a local researcher, which suited local circumstances. I was astounded to learn that the World Bank had insisted on the purchase of compactor trucks, which required the bulldozing of shacks to make access roads. In addition, the vehicles were often out of action because the local people did not have the parts or expertise to repair them.

Policy makers at all levels have to learn to allow communities to develop through the dynamics of local engagement. Such process takes account of local resources, culture and capability. Western 'experts' can find this particularly hard because, although we can share technology and offer information, it is vital that we do not dictate terms and force development that mimics the solutions we have at home.

Community engagement can descend into conflict and argument, of course, but just as easily it can result in people vying to win status by showing how much they can do for or give back to the community. Such behaviour conflicts with the assumptions of capitalist economics, but it is deeply rooted in human behaviour because, over thousands of years, such selfish altruism has succeeded in making humans the dominant species on the planet.

Human-Scale Towns and Cities

In a sustainable society, people should live in human-scale, close-knit communities in which the development of sustainable policy is a natural outcome. This is not the case in many contemporary towns and cities, where people live in segregated private compounds, do not know even their close neighbours and make every journey by car. The loss of community values can be extreme, leading to people believing that their needs are demands that must be met, that everything is available for a price and that community cohesion is provided by the rule of law. I believe this to be a damning indictment of modern society, though many people see nothing wrong with the situation I describe. They would claim that whatever they need, society must provide; whatever they want, should be on sale; and that it is the responsibility of the police to maintain a secure and stable society. These attitudes come from the era of plenty when people were encouraged to take responsibility only for themselves. In the era of diminishing resources, such selfish attitudes will be exposed and the people who fail to adjust will be on the receiving end of a backlash from those who are the first to suffer. The cohesion of society could fall apart unless society changes direction.

Anyone who believes they can stay aloof from the need to change behaviour in response to risks to the environment and diminishing resources is deluded. To be safe, we all need to belong to a sustainable community. Such communities are more about people than infrastructure: people who share common aims and aspirations and work together to organize their affairs. In such communities, problems can be overcome and the setting of sustainable policy becomes feasible.

The concept of human-scale communities is not new, but in the twentieth century its importance was forgotten as pure economics influenced policy making. In the twenty-first century, the foundations of community need to be rebuilt on human interaction and human relationships, informed by scientific analysis and logical deduction. The physical form of the communities will be fundamentally determined by the logic of spatial distribution.

Spatial Distribution

The spatial relationship between objects and places in a sustainable world is important. In the globalized world of fast, cheap transport, the cost of getting from A to B was small, so we stopped worrying too much about the distances involved. Transportation costs were only a small factor in decisions on where to locate factories, housing, shopping malls and so on. A reliance on transportation – to an extent that is only possible if fuel is cheap –has been allowed to shape communities to the detriment of quality of life. I use a simple spatial analysis below to draw out the issues. The reader is encouraged to go along with my approximate assumptions so that I am able to explain the core logic succinctly.

Some simple metrics can be used to define the spatial distribution of urban services and facilities. Let us assume that walking is acceptable for journeys of up to about 600 m and journeys by bicycle acceptable to about 4 km. This means that an area of approximately 1 sq km around our place of residence is within easy walking distance. An area of approximately 50 sq km is within easy cycling distance. These figures need to be related to urban density and the numbers of people required for support services to be viable.

I will use a rule of thumb that access to a population of 3,000 or more people[1] is required for a community to support an urban centre with a full range of services and shops to cover basic day-to-day needs. On this basis, if walking is the prime means of shopping, a population density of 3,000 people per sq km is needed to support an urban centre. If cycling is predominant, a population density of 60 people per sq km would suffice.

Another aspect to consider is the provision of rail and bus services to link communities. In the rule of thumb provided from an Australian perspective by Moran (2006),[2] rail-based systems require a population density of 40,000 people per sq km to be commercially viable and express bus systems need 25,000 people per sq km. These figures make assumptions about the level of take-up in a society where car ownership is widespread. Adopting policy to discourage car ownership could make lower densities viable.

A sustainable society needs local services within walking or cycling distance and access to buses and trains. On the rough assumptions above, our

1 To arrive at the figure of 3,000, I selected the population of Pangbourne, Berkshire, UK. This is a small vibrant town with a range of services, including a supermarket, butcher, greengrocer, post office, flower shop, clothes shop and a selection of pubs and restaurants. Note that if policy turned against the large out-of-town superstores, communities smaller than this could support a full range of shops and services.

2 Moran puts forward these rule-of-thumb figures but it is worth noting that he also argues that public transport can be delivered at lower densities. I use the rule of thumb to make a general point without claiming that these are definitive figures.

communities should have a density of between a minimum of 60 people per sq km up to an ideal of over 40,000 people per sq km.

A study of housing density across the north-central United States found that in 1940 population was concentrated in dense urban centres, but that by 2000 there had been a large expansion of suburbs with densities of two to 16 dwellings per sq km (Stewart et al. 2003). Such low-density living is reliant on the car. A transport mix including walking, cycling, buses and trains is not feasible. In the UK, denser suburbs of typically 5,000 dwellings per sq km were built over the same period (Kochan 2007). Although better than in the United States, this density is also not enough to support a good sustainable mix of transport choices.

In moving towards sustainable communities in the UK, housing models are being developed with densities of up to 12,000 dwellings per sq km but which retain suburban qualities (Kochan 2007). Assuming an average occupancy of 2.5 people per dwelling, this provides sufficient density to support good public transport provision. Housing design guidance for London is being developed for 'superdensity' developments of 15,000 to 50,000 sustainable family dwellings per sq km (Design for Homes 2007). This is the direction that city design must take.

The logic of spatial distribution leads to the conclusion that denser living is required within a sustainable society. People have to accept a smaller piece of the Earth as their private space. This conflicts with the widely held aspiration in the West for a large house and large garden. If we insist on this way of life, we have no choice but to be enslaved to the car. The way forward is to reframe our aspirations from living in large private spaces to living in communities with high-quality shared space.

Public and Private Space

By insisting on indulging our desire for large amounts of private space, we undermine the possibility of constructing sustainable communities. A compromise is required between reserving space for our exclusive use and ensuring that there is space shared with others in order to be able to grow a strong and cohesive community. Private space can be very small, without compromising quality of life, if there are high-quality shared spaces and facilities.

This is not a rich versus poor dilemma. Both rich and poor can benefit from the concept of quality shared space. Rich people may prefer a gated community with controlled access and within which there is a swimming pool, tennis court and other facilities. These facilities are shared with the other members of the community, rather than wastefully providing one per household and isolating their owners' lifestyles. Poorer communities will be more reliant on public facilities, but the concept is identical. Poor communities may not legally own the facilities, but city planners should design neighbourhoods in such a way that makes it easy for communities to take ownership of 'their' space. Close oversight by the community should be encouraged to ensure that community spaces are maintained for the collective good.

The acceptance of less private space in order to be able to support more cohesive communities will require a change of attitudes and expectations. The aspiration for a large house and private garden is many people's dream. Changes in attitude will come about as the benefits of high-quality communities become evident. In addition to persuading people that less private space is acceptable, the other big cultural shift required is to reduce the status of the car within the built environment.

Reversing 'Car-Centric' City Design

Moving away from the megametropolises brought on by twentieth-century suburbanization (WBCSD 2001) will be a tough policy shift to make. Policies to free cities from dominance by the car are well defined (Crawford 2000), but not widely known or accepted.

In the developed world, the car holds a revered status and owning one is seen as an unalienable right. In the poorer countries, owning a car is one of the main aspirations to be satisfied as disposable income grows. It can be assumed that our engineers will succeed in designing and building clean, recyclable cars running on renewable fuel. This will remove the obvious problem of pollution caused by cars. This will not be enough to fix the problems attributed to cars. The addiction to the car leads to sprawling suburbs and undermines policy to develop cohesive communities (McManners 2008: 107–111).

I presented a paper on removing cars from urban life at the London School of Economics in 2007.[3] I argued that the developed nations had made a mistake in allowing 'car-centric' urban design to dominate towns and cities. It will take time and effort to alter the infrastructure of cities so that they are built around the needs of people. The sooner the developed world acknowledges the mistake, the sooner the transformation can begin. I then extended the argument to promote policy for the developing nations to bounce past our mistakes and build cities around people.

My paper received support from a number of people from the developing world, but the US officials from the World Bank were adamant that the return on investment in road building was better than the dense developments supported by public transport that I proposed. The US model of urban design appeared to be firmly entrenched and my argument was not going to change it.

There are US commentators who are critical of the US urban model, such as James Kunstler (1996), but he is regarded as a maverick and is not given serious attention by government policy makers. The United States has a huge problem because the 'car-centric' model has been implemented universally to a high standard and it works very well. It is hard for the United States to see that there are better alternatives. Kunstler predicts it will take the crisis of oil supplies running low for the US urban model to collapse (2005). Persuading the United States to make a deliberate choice to shift to a different model will be hard as it requires completely redesigning their urban places. I hope that the United States starts the process in good time to avoid Kunstler's vision becoming a reality. There are hopeful signs of a sensible dialogue emerging in the United States with the formation of CarFree City USA (2009)[4] in California in 2003 and in recent publications such as *Resilient Cities: Responding to Peak Oil and Climate Change* (Newman et al. 2009).

Instead of berating the United States, we should understand the depth of the US problem. Not only are cities built in a way that makes car use unavoidable, but also the car is an icon of American culture. US politicians face a huge challenge to find a sustainable way to deal with US car dependency. How they do this is an internal matter for the United States.

3 'Cities for People: Removing Cars from Urban Life', a paper presented at the World Institute for Development Economics Research of the United Nations University (UNU-WIDER) project workshop on 'Beyond the Tipping Point: Development in an Urban World' held at the London School of Economics and Political Science, 19–20 October 2007 (McManners 2009: 129–45).

4 CarFree City USA is a non-profit organization founded in 2003.

As the United States debates its own problem, other countries must not be sucked into the same honey trap. The car is convenient transport, and also an object of pride and symbol of status for many owners. Marketing executives of car corporations work hard to indentify cars with status, wealth and even libido. This message is reaching developing countries where the aspiration to own a car becomes a high priority as people get richer. Not enough is done to expose the scourge of many thousands of cars undermining community life and taking space that could be put to better use. This is the true message that should be coming from the developed world to counter the car adverts in glossy magazines.

Officials in influential positions in international organizations who have grown up within the car culture and have not yet been converted to thinking about policy in sustainable terms are of particular concern.

I admire the smooth lines and good engineering of a fast car as much as anyone, but I want to live in communities where cars are not essential for everyday tasks and people can get around safely on foot and by bicycle. My views stem from a wish to improve quality of life, but I know from bitter experience how deep the resistance can be from people who do not share my view of what constitutes better cities and towns.

As the difficult task of changing attitudes starts to show progress, economics will have to adjust to support 'people-centric' urban design. Twentieth-century suburbs will have to be dismantled to be able to build communities fit for the twenty-first century.

Economic Levers of Change

Building cohesive twenty-first century communities will be a complex process of attitude shift and expectation management. It will also be fraught with political difficulties. To support this process, I provide two examples of simple economic levers that could be applied.

To unwind suburbanization, a flat tax could be levied on all urban areas based on the area of each land parcel. This is a particular application of the urban eco-balance tax introduced in Chapter 5. Such a tax would be easy to administer by piggybacking it on other property taxes, and it would be easy to enforce with the ultimate sanction of taking ownership of the site. Such a tax would be not based on the value of a particular site but its area. For a valuable site at the heart of a town or city, the additional tax would not be significant. For a large low-value site on the periphery, the tax would be significant and would be a disincentive to low-density development.

When I first proposed this tax (McManners 2008: 198–9), I was concerned with both a retreat from the policy of expanding suburbs and the need to retain land for nature. To address the latter aim, part of the tax income would be paid to the owners of land kept for nature. I went further to suggest that suburbs could be redesigned as compact sustainable developments with surrounding areas returned to nature, thereby offering a tax-neutral solution in response to this particular tax lever (as well as improved quality of life for residents).

The other main change to support my vision of sustainable communities is the introduction of measures to reduce car dependency. Increasing fuel costs and taxes on fuel will lead in this direction, but more can be done. All the land devoted to car infrastructure is normally regarded as public land kept for the public good, so no rent is charged. If a city is designed on the assumption that everyone owns a car, free roads are a public benefit. By shifting to a society in which cars are optional, roads are no longer a universal public benefit so subsidies (direct or indirect) can no longer be justified. Car drivers should be charged an access charge that fully reflects the value of the land taken up by roads. If the income is then invested in public transport, such policy can eliminate the concept of universal car ownership without being seen as unfair.

Instead of the automatic response of jumping into the car, people will have to make a conscious choice over how to travel and where to live to have access to good public transport. Breaking the link between people and individually owned private cars is the key to reducing car dependency. This means that taxis, hire cars and minibuses should be treated leniently so that people have confidence in living without a personal car. There will be occasions when the point-to-point convenience of the car is hard to replicate, so easy access to affordable and energy-efficient hire cars will help to persuade people that a personal car is unnecessary. This will be particularly important in the early years, before the full benefits of lower car dependency flow back into society.

These are examples of straightforward economic methods that can be used to help shape human communities. Twenty-first-century communities can be different to the sprawling urban suburbs of twentieth-century design. When communities are drawn together to discuss the future, under pressure from escalating fuel costs, and with governments under pressure to reduce carbon emissions, the circumstances will be right to change the status quo. The final push will come when a growing number of communities demonstrate in a real-world context that sustainable living works and works well.

Twenty-First-Century Communities

Since the Industrial Revolution, people have migrated from the countryside into cities. They have been pulled by opportunities to work in offices and factories, and pushed off the land by the growth of industrialized agriculture. The landscape has shifted towards a checkerboard of monoculture agriculture and urban development, with nature squeezed out. As the demand for agricultural output increases and city design shifts to higher-density urban living, there is a danger that the delineation of urban and agricultural spaces becomes yet more distinct, with nature confined to national parks and nature reserves.

The shift to true sustainable communities will follow a different path. Cities freed from air pollution will become locations to grow food on balconies and rooftops. Such food could not be fresher or more sustainable, because it uses space that would otherwise be wasted and reduces the need for transportation. As agriculture becomes more sustainable, considerable thought, care and effort will be required on farms to maintain similar yields. The use of fertilizer and insecticides can be reduced by careful management of farms as extensions of the natural habitat. Integrated mixed agriculture can also reduce the need for energy-intensive machinery. This is not to suggest replacing machinery with people doing soul-destroying, back-breaking work. It suggests that clever twenty-first-century agriculture requires more human engagement.

As people are brought into a close relationship with nature, the value of ecosystems will be appreciated first-hand, making it easier to reach the compromises required so that nature and human society can coexist in harmony.

Agriculture will be brought into the cities and some city folk will shift back to small agricultural communities. Nature will be invited back across the whole range of human space, providing the habitats and corridors between them that are needed for species to have a fighting chance of surviving the impacts of climate change.

The most valuable space within communities is that which is open to the sky. We can always build up or dig down, but the open-to-sky space is finite. In twenty-first-century cities, this will be carefully conserved for people (to sit, walk or cycle), for agriculture or for capturing energy from sunlight. In densely populated cities, we will be able to afford to put all forms of mechanized transport underground.

Outside cities, there will be a rash of small communities, not just for people working in agriculture, but also for knowledge workers in small offices who are connected with their colleagues by high-capacity data links. The economics will lead to the retention of a surface road network, but the space that roads take up will be better used. We will see a return to the tree-lined avenues that were once common, for example in France, where the sunlight that would otherwise be wasted is used to grow timber.

The changes I champion for twenty-first-century living will be seen by some as a killjoy process of taking away the good life of cars and grand suburbia. The reality will be much better communities that reinforce the integrity of the environment and which deliver more satisfying engagement for people with each other and with nature. This is not a retrograde step; humankind will continue to make advances in technology and knowledge to further improve quality of life. Twenty-first-century society will be highly advanced and it will operate at human scale. This is the key to success for the communities that will thrive in the twenty-first century.

15

Sustainable By Design

All prudent princes ought to have regard not only to present troubles, but also future ones, for which they must prepare with every energy, because, when foreseen, it is easy to remedy them; but if you wait until they approach, the medicine is no longer in time because the malady has become incurable

Niccolò Machiavelli, The Prince, 1515

Prior to the Industrial Revolution, human society was sustainable by default. Human activities were at small scale compared with the immensity of the Earth, so that whatever humankind did, the stability of the global ecosystem was not put at risk. Communities could collapse as certain cultures and behaviours were found to be flawed. Whole civilizations could disintegrate, consumed by corruption, complacency or any number of failings that afflict human societies. There would always be another tribe to take up the running for humanity to continue.

The modern globalized world is different, and people are slow to understand the significance. Humans now have the power make irreversible changes to the ecosystem. Modern human society is so advanced and has such capability that it is feasible for humans to damage the ecosystem to the extent that it would no longer be able to support human life. That is the extent of the power we have acquired. We must – and I am optimistic that we can – find a way to wield this power responsibly.

Concurrently with expanding the scope and scale of technological capabilities, human society has become interconnected and interdependent as never before. This gives the illusion of greater stability. A localized problem is soon resolved by importing supplies from the world commodity markets or borrowing capital from world financial markets. The possibility of local collapse in a society or an economy is lessened. The reliance on global markets within a framework that looks increasingly like the emergence of a global society is dangerous. There has never been, and never will be, a perfect human society. In a globalized society, in which every economy and country is intertwined, any flaw will, over time, be hugely magnified. The price of the medium-term stability that globalization can bring is the occasional dramatic collapse. This is not something that can be prevented: it is a feature of any human system, and when that system is all encompassing there will be no escape from the fallout.

Modern human society is no longer sustainable by default, so it must be made sustainable by design. This 'new' design framework is proximization, as I explain and describe in this book. It is the natural way to run world society. It is the way society would be run if a narrow focus on economic outcomes had not intruded. The proximization policy framework relegates economics from being on a pedestal to being a mechanism of implementation. Economic policy remains a powerful tool, but the focus should be on social provision and environmental protection, using levels of rigour so far reserved for management of the economy.

Changing the Economic Levers

The economic levers that can tip society towards a sustainable future are different to the policies of economic globalization that have been used widely over the last three decades. Policies such as free trade and free capital flows have been largely successful within the narrow confines of delivering economic outcomes. While society continues to measure success with economic measures, such as GDP, wider implications, such as increasing stress on the environment, are largely ignored. Economists get around this obvious omission by calling such effects 'externalities'. This convenient pigeon-holing of inconvenient consequences can no longer be justified.

The Club of Rome warned in 1972 that there is a limit to growth (Meadows et al. 1972), but the warning was ignored. Over three decades later, the world economy has grown to $65 trillion in 2007 (IMF 2008). Economic growth was particularly strong from 2000 to 2007, leading into the recession of 2008. The world economy grew at an average of 4 per cent, led by record growth of 6.5 per cent in the economies of the developing world (World Bank 2009b). The experience of the last few decades has fooled us into believing that we can carry on with business as usual indefinitely. This is a denial of logic. Recent research indicates that the warnings by the Club of Rome were correct (see Chapter 2). The fact that there is a limit to growth is hard to deny; the time has come to change the economic levers with which society is steered.

Following the financial crisis of 2008, world leaders tried to boost economies to recapture growth. They were still locked into the policy of measuring success using the metric of GDP, despite the evidence that this will not work over the long term. It is understandable that, in the immediate urgency of financial crisis, politicians reach for familiar policies. As short-term stability returns, policy makers should be looking for the long-term sustainable solution.

The proximization policy framework aligns economic levers with the needs of sustainable society. People do not see systematic global problems as their personal business. The key to the success of the proximization framework is that it brings the challenges facing human society within the sphere of national control, with decisions based on local circumstances. However, proximization is not synonymous with isolation and is not a policy of localizing everything. There will still be global trade, but at lower volumes, based on real needs and sustainable ways of satisfying these needs.

The Proximization Framework

The first principle of proximization is that decision making is based on sustainability – that is, the delivery of social outcomes through policies that reinforce the ecosystem balanced with economic viability. Economic efficiency is one factor, but not the overruling factor. For example, there will be situations when quality of life can be improved through increased levels of human engagement (jobs) and low levels of virgin commodity flows (clever design and complex intelligent processes). These may not be the most efficient solutions in terms of conventional economics. A narrow focus on economics can lead to 'fixing' such situations to deliver a better return. Such corrective action may be supported by the treasury ministry and the accountants, but requires the offloading of the externalities – such as social outcomes – to another government department. Environmental degradation is another externality that is given less attention than it deserves, thus handing the consequences on to future generations. Adopting the proximization policy framework makes it much less likely that decisions which undermine sustainable society are taken through the inappropriate application of economic efficiency measures.

The second principle of proximization is subsidiarity. People engaging in communities discuss, negotiate, argue and make the compromises required to balance competing issues. Decisions should, therefore, be taken at the lowest possible level in society. This requires restraint from higher authority, particularly with regard to international relations. Governments must resist the temptation to impose ideology on other nations when it risks undermining their capability to develop a sustainable society that suits their circumstances. Control at higher levels should be restricted to only those policy areas where there is a clear benefit. Of course, discussions take place within a framework of regulations, particularly with regard to the environment. The setting of this framework needs robust forward-looking action from national governments. This leads into the third principle of proximization.

The third principle is the primacy of the state. The state has a clear remit to coordinate policy across society, the environment and the economy. This is the level of control in world society where the ability to take action is greatest. The state has formal power over legislation, borders and the rights to extract natural resources. The nation state also has the soft power of drawing on loyalty and national pride. Considerable personal disadvantage can be tolerated in times of crisis when it is seen to be for the national good in a country with a strong culture and deep-rooted shared values.

The fourth principle of proximization is the use of market economics to mobilize the invisible hand of the market to deliver sustainable outcomes. The context within which market forces are used is vital and is different to the blind faith that has taken hold in the financial and equity markets in recent decades. Sustainable markets will require close oversight and tight regulatory controls. Decisions over desirable sustainable outcomes will set the context within which the market operates and around which the regulatory framework will be crafted. Where the sustainable solution is specific and obvious, direct precise economic levers, such as targeted taxation set over a number of years, will provide the basis for a sound economic case for the required investment. Where solutions are not obvious – which will be a common situation throughout the Sustainable Revolution – then market mechanisms have the required flexibility and neutrality to reach a solution. Competing technologies, methods and processes can fight it out in the market, thus supporting the development of innovative solutions.

A Retreat from Global Markets

The markets designed to deliver sustainable outcomes will be national, with few exceptions. The danger of relying on global markets is that global governance is too weak for effective enforcement. Unconstrained global markets focused on economic outcomes have driven a wedge between economic policy and environmental policy. This is not a situation that should be perpetuated. Unless global governance can be dramatically improved (an unlikely eventuality), global markets should be resisted. Attempts to improve world governance should take place, of course, to align international social, environmental and economic frameworks, but care should be taken not to expect too much. Countries are not easily persuaded to compromise national self-interest in order to achieve the sort of equitable global deals required to justify relying on global governance to protect the global ecosystem.

Policy makers should seek to establish linkages between national markets when sustainability can be designed into the agreement in a way that is enforceable. There will also be a role for regional markets, such as the EU, where the interests and aims of a number of countries are closely aligned. The EU leads the world in the development of environmental policy, such as carbon trading and recycling. These initiatives could evolve into a robust and workable sustainable policy framework for Europe. This will require strengthened European governance because integrating the open European market with sustainable policy will require some difficult decisions in which national interests will have to take second place to European sustainable policy. The EU Emissions Trading Scheme (ETS) is an example which has the potential to lead Europe away from fossil-fuel dependency. Looking forward to the time when the price of carbon is high, and the economic cost significant, EU politicians will have to find the political strength and unity to hold firm to the carbon cap despite attempts to secure national dispensations.

The Short-Term Real-World Context

The aftermath of the financial crash of 2008 generated considerable discussion about the need for better regulation of financial markets. Frank Ackerman (2008) wrote, 'The market may be the engine of a socially directed economy, indispensable for forward motion. There are limits, however, to its capabilities: it cannot change its own flat tires; and if we let it steer, we are sure to hit the wall again.' Inevitably, the focus is on fixing the existing system. It is not possible to close the markets and shut down the economy while a better system is being designed and then rolled out. The economists have a lot of work ahead to analyze the causes and come up with solutions. In moving beyond the crisis, it is important to set the high-level context of how society should be run to give economists the guidance they need.

For the immediate future beyond 2010, policy makers have focused on lifting the world economy out of the worst recession since the Great Depression of the 1930s. Policy makers have been looking back at this era for inspiration. US President Franklin Roosevelt's action was to launch the New Deal. This entailed a strong government role in economic planning and a series of stimulus packages launched between 1933 and 1938. It created jobs through ambitious government programmes that included the construction of roads, dams and schools. In 2009, Barack Obama proposed a New Green Deal to deal with the economic and climate crisis together.

The idea that a New Green Deal could address the 'triple crunch' of a financial crisis, accelerating climate change and soaring energy prices as oil production peaks, was first presented in a report by the new economics foundation in London in July 2008 (Elliot et al.) In the US, a report compiled by the US organization World Watch for the Heinrich Böll Foundation called for wide-ranging decisive action (French et al. 2009).The UN took up the theme with its Global Green New Deal (GGND) presented to the G20 meeting in London in April 2009. The UN set three objectives for the GGND (Barbier 2009):

- Revive the world economy, create employment opportunities and protect vulnerable groups.
- Reduce carbon dependency, ecosystem degradation and water scarcity.
- Further the Millennium Development Goal of ending extreme world poverty by 2015.

Through 2009, efforts to implement a Green New Deal around the world were disappointing. Economic stimulus was given far more attention than the green policy initiatives. To rescue this initiative and secure its place in history alongside Roosevelt's New Deal, there will need to be much more radical action to deliver real progress towards sustainable outcomes, such as a zero-fossil-carbon economy.

Another important issue at the end of the first decade of the twenty-first century is the expansion of the Chinese economy and China's ascendency in the new world order. This highly populous country has grown from a slumbering giant based on an agrarian economy to become the workshop of the world and to surpass the United States as the world's biggest emitter of carbon dioxide. This is a huge problem for the world. China is consuming at a rate over double its ecological capacity (Global Footprint Network 2008).

The United States is often accused of being the world's most unsustainable society, but this can change. The United States could balance its consumption with its ecological capacity if it halved consumption to match the average level of consumption in Europe (Chapter 6). From a European perspective, this would appear to be feasible. However, consumption in China is a relatively low 2.1 gha per person. If China aspires to European levels of consumption, its deficit will rocket and it will have to scour the world for supplies to satisfy consumption. The West is slow to understand that encouraging progress in China along Western lines is a policy own goal with negative outcomes for the world.

It would be better for the planet if China found a Chinese path to sustainable living. It is not for those living in the West to say how. History may provide some clues. Two hundred years ago, leading parts of Europe and China had reached similar levels of development. *The Economist* (2009) reporting the work of Martin Jacques (2009), argues that the factors that underpinned Europe's success relative to China in the nineteenth century were its ability to wage war, possession of colonies run by slave labour and the promotion of merchant classes into the elite, encouraging European rulers to promote capitalism. Meanwhile, China remained attached to Confucian values of harmony and social equality. With a history such as this, Europe should be reticent about claiming that the West's values are superior. The time has come to pull back from encouraging reform in China along Western lines and instead allow China to find a sustainable harmony in its own way. The prime reason for adopting such policy would be a selfish desire to reduce the Chinese impact on the planet, but the West would also have the opportunity to observe if there are Eastern values or behaviours that the West might usefully adopt.

Globalization Reversed

Further moves towards a globalized and homogenized world should be shelved and changes should be made to reverse out of the policy cul-de-sac in which world society has become trapped. To embark on the next stage of human progress, innovation, drive and determination will be required; but this will be according to a different paradigm. The changes will come from the bottom up as countries put policy in place to ensure their own future with some protection from the vagaries of the world market. There could be a rearguard action by free-traders championing open markets – while surreptitiously protecting their own interests. Alternatively, there could be widespread acknowledgement that the world needs the changes described in this book so that the shift to proximization takes place in a reasonable orderly manner. Whatever route the world takes, the destination has to be a sustainable world society – anything less will mean that we have failed the generations of people yet to enter this world. Human nature means that the path is likely to be circuitous and will include a number of policy experiments that fail before we are able to settle on a stable and secure future. It would be better if this came about through design, rather than chance and experiment.

A New World Order

A new world order is required to solve the imbalances that have arisen between human needs and the stability of the ecosystem. The progress of recent decades has been fuelled by economic globalization. We want to retain these advances, but it has become clear that consumption and expansion will strip the Earth bare. The time has come to move forward according to a new paradigm. The changes involved are a massive U-turn in policy, but the effects of the new paradigm can be applied rapidly and effectively because the policy of proximization fits with the way that human society evolves – if economists are kept at a distance from decision making.

Proximization is a natural and stable structure for world society. The variety in societies and economies provides the diversity required for macro stability, much as an ecosystem needs a broad gene pool. It is natural that societies try to apply local control to protect their resources and ensure their economies are working for the benefit of their people. This is why the ideology of economic globalization has been so hard to sell to many countries. The simple action of backtracking from the ideology of economic globalization would be enough to initiate progress towards adopting proximization.

Orthodox economists have championed free trade and free markets, and IMF officials have forced such policies on to reluctant countries, because of the economic benefits. The economists have facts and figures to back up their advice. There are many positive outcomes that show up in the figures for trade and GDP. It is a fact that economic globalization delivers growth in GDP for all participating countries. If GDP is the measure of progress, economic globalization is the correct policy.

GDP (or its predecessor Gross National Product (GNP)) became widely used following the Second World War. Since then, it has grown in status as the prime metric for economic performance and has become a proxy for progress. The meticulous work of Simon Kuznets (1941) in *National Income and Its Composition, 1919–1938* was significant in defining the basis of GNP. Ironically, in an earlier report, Kuznets wrote, 'The welfare of a nation can scarcely be inferred from a measure of national income.' (Kuznets 1934).

GDP may be a good measure of total economic activity but it should never have been used as a measure of human progress. This has trapped policy makers into focusing on increasing GDP rather than the much more complex task of increasing, to use Kuznets' words, the 'welfare of a nation'.

Initially, sustainable policy seems to involve restrictions, and attracts resistance. This knee-jerk reaction has to be overcome in order to be able to appreciate a society in which there is rich depth in social engagement and considerable enjoyment from living in tune with nature. Pleasant surroundings need not be a perk only available to the rich, but should permeate across society. Economic outcomes will no longer be the prime purpose driving behaviour. Controls will be brought in to stifle the casino culture that has afflicted the financial markets. Gambling in the financial markets seemed a reasonable activity when success was measured by financial return; when juggling policy to maximize financial return means gambling with the future of the ecosystem, this is not acceptable. Economic policy should be stable, predictable and rather dull.

The current generation of world leaders will find it hard to adopt different ways of formulating policy and making decisions. In summer 2009, I was a delegate at the World Forum on Enterprise and the Environment.[1] Amongst what was otherwise an excellent discussion on responses to climate change, there was one session that fell flat. The session was introduced by an economist and university professor who started by laying down some parameters for the discussion. First, we could not propose anything that may be a barrier to free trade. Second, any proposal must not cost more. These principles were

1 World Forum on Enterprise and Environment, 5–7 July 2009, organized by the Smith School of Enterprise and the Environment, Oxford University, UK.

deep-rooted in his understanding of orthodox economics. These preconditions also ruled out all but business-as-usual options. This is the message that governments get from their economic advisors and which is taught to students of economics. The message is hard to counter because, in the narrow confines of pure conventional economics, it is the correct answer. It takes courage to respond by saying that economics is an inferior discipline to social provision and environmental protection. The best solution may not be the most economic solution. We know this to be so in the way we live our own lives, but we have to fight to translate these wider values into the way that governments run society. The best solution should be judged on social outcomes that fit within sound environmental stewardship, without departing from the discipline of sound financial planning and control. This is a complete reversal of priorities compared with selecting the most economic option, and then applying an environmental and social impact assessment.

People will have to be persuaded and convinced that a sustainable society is both necessary and can deliver an improvement in the quality of life. It would be disingenuous to claim it will be better in every way, because it will require reining back on material consumption and treading rather more lightly on the planet. Parts of the modern lifestyle, which we now take for granted, will no longer fit. Buying cheap products to use for a short time and then throw away will no longer be an option. Only quality products built to last and designed for total recycling will be on sale, and they will cost more. Computers will be designed so that key components, such as processor and memory, are easy to replace so they can be upgraded rather than chucked out as junk after a handful of years. Those of us campaigning for a sustainable society see clearly that this is a preferable option, but the initial reaction of people conditioned to search out the cheapest option is to object to the increase in cost.

When sustainable policy making is ingrained in society, sustainable choices will become instinctive. The dichotomy that can exist between economic and sustainable outcomes will disappear. Partly this will be because we have adapted economics to serve society better, and partly it will be because society has become sustainable by design. When teachers teach respect for the planet in primary schools, when universities explain the need for a sustainable balance as the basis of society and when business schools embrace green economics, then the world will be led by people capable of steering the world on to a safer path.

People should be encouraged to strive for quality of life. Achievement should not be measured by a template presented in US or European soap operas or defined by targets set by mandarins in the UN or other international agencies, but through finding a balance in people's lives and in their communities that makes the best of local resources and circumstances.

Adam Smith would have been horrified how his theory of the division of labour within a cohesive society had been subverted to justify the excesses of globalization. The invisible hand needs to be brought under the control that Adam Smith took for granted in the close-knit localized society of his day. He was an intelligent and compassionate person who observed efficient and effective ways to run society. I believe that he would support the proposals made in this book.

It is clear that, for long-term sustainability, the world needs escalating taxes on fossil fuels, robust protection of natural assets and real measures to tackle population growth. To make progress, we must expose the unsustainable nature of the current policy framework and, in tandem, propose and sell a new framework that can deliver long-term stability.

The new world order will pull back from the worship of consumerism to strive for success as defined by happiness, health and status derived from making a contribution to the community. Less intense resource consumption will take the pressure off the ecosystem and will start to alleviate the big environmental challenges of our age, such as climate change. It will take time for people to accept a different set of metrics and change deep-rooted aspirations. Time is needed, but the time available is short. Already there are environmental consequences that can no longer be stopped. As temperatures increase, many glaciers will disappear and global sea level will rise – it is only a question of by how much. To prevent more serious consequences, the process of change in society must start soon. People must move from talk to action.

Political Leadership

The challenges of the transformation to make society sustainable require bold and rapid action. This demands the sort of leadership normally only seen in times of war. The struggle for diminishing resources will indeed lead to war in many places, but it would be a sad reflection on humanity if substantive action were delayed until that point.

In the political world of spin, focus groups and opinion polls, the pace of progress is glacial. If each decision requires majority support before implementation, there is little chance of succeeding. The systematic changes required will only show benefit when there is success on a number of fronts. Until then, there will be changes that appear to cause hardship and to be restrictions of freedom: not the real hardship and sacrifice of war, but real enough to be politically difficult.

Politicians need the courage to take a principled approach. An example is action to reduce CO_2 emissions. To date, considerable discussion has taken place about carbon trading and subsidies for renewable energy. This is a politically easy route, but it skirts around the main issue. Fossil-fuel prices have to be set on an unequivocal and steep upwards trajectory to give clear signals to the economy. Politicians must accept this as fact. The explanation to the public could focus on climate change or energy security. The latter may be easier to use politically but, whatever political argument is used, the core fact that fossil-fuel prices will escalate must not be obscured.

Ideally, the core policies of sustainability require cross-party support so that key elements are not open to political wrangling. There are some actions that must take place whatever opinions people have, such as the need to pull back from reliance on fossil fuel. Western leaders may look to China with envy if the authoritarian government there turns its attention towards the task of building a sustainable society. The West may not want to copy Chinese methods, but Western democracy will need to be reinforced, as outlined in Chapter 12, to be able to take substantive action.

Business Leadership

A recurring theme in my writing is the concept of using business as the primary agent for change. Of all the principal actors in society, business is able to move the most quickly. It is not constrained by the need to win consensus nor held back by lack of funding. When a good business opportunity emerges, an entrepreneur will spot it and persuade venture capitalists to provide the funding. Unlike government plans, business plans backed by venture capitalists do not need to offer solid and certain returns. Venture capitalists can tolerate a number of failures in order to own stakes in the businesses that succeed.

People outside the business community can reinforce its role as the primary agent for change. Business needs customers, so customers have power. Customers exert that power by where they choose to spend their money. Society – that is customers – has to be persuaded of the need to act before the power of business can be fully mobilized. As soon as society reaches the tipping point where 'green' considerations will be de rigueur in all purchasing decisions, then business will respond.

Within business schools, it is vital that the teaching includes sustainability and the concepts of a green economy. Management teaching has become caught in a rut through a narrow short-term focus on increasing shareholder value and almost unquestioning commitment to the concepts of free trade and open markets. The deep-seated assumptions underlying teaching have to be challenged in order to open up research and start the dialogue for the next wave of innovation. The discipline of corporate responsibly is gaining traction, but its concepts tend to be directed towards enhancing reputation rather than feeding into core business strategy. Much more progress is needed in this area, but when this progress has been achieved, it will have huge potential to increase society's ability to be sustainable by design.

NGO Leadership

There is a proliferation of NGOs focusing on every conceivable issue and concern. In a complex world, a wide variety of NGOs pursuing different agendas can have the effect of neutralizing each other's efforts. International agreements brokered by the UN under the influence of NGOs and other lobbyists often end up with a plethora of aspirational aims instead of tightly focused priorities.

For example, one NGO can champion industrial development to raise living standards and reduce poverty, while another is seeking to protect the natural environment from exploitation. One NGO can seek to feed hungry people, undermining the efforts of other NGOs that are seeking to build community and self-reliance. NGOs navigate through these dilemmas with the very best of intentions and there is considerable inter-NGO dialogue, but I would like to see overarching priorities emerge based on long-term sustainability.

It is clear that sustainability should underpin government decisions and be understood by business. NGOs should also bring sustainable thinking into their work to ensure that their particular concern is handled in way that makes positive linkages with other areas of policy. Every campaign and project should conform to a robust model of sustainability, being careful to avoid the trap of assuming that sustainability must be linked with development. Development must always be sustainable; sustainable policy does not need to include development. Focusing on development is not appropriate policy for the world unless the definition of development is changed to focus on broad social and environmental outcomes (see Chapter 2).

Reaching the Tipping Point

Politicians, business leaders and NGOs have their hands tied by the need for popular support. They cannot act until the majority of a particular population are willing to commit to sustainable living. The idea that the world's countries could act in concert to shift to sustainable living is idealism that has no place in the real world, except to act as a smokescreen for delay and inaction. Proximization encourages a country-by-country approach to find sustainable policies that are feasible and politically acceptable. As the population of a country tips in favour of sustainability, it will start to concentrate on securing a safe future for its society. As other countries observe this process, they may follow. This could be the catalyst that sets off a domino effect leading to the sustainable world that humans need to continue to thrive.

There will be resistance to the shift to sustainable policy because of the short-term downside. People live in the here and now. It may be necessary to wait until the world's problems are so severe that they are having a direct effect on the majority of people before change is accepted. This would be a sad reflection on humanity. It is time humans took the next step in progress to learn to apply the advanced capabilities and technologies we have developed. We must find a way, soon, to set society on to a sustainable path. If we cannot, or will not, find a new direction, then modern civilization will add a new characteristic to humankind's defining qualities. Not only will we be the most intelligent, most successful and most populous species ever to inhabit the Earth, we will also be the most self-destructive.

Epilogue

Whilst this book has been going through production it has already generated debate. I am grateful for the enthusiastic support I have received and also for the warning that my pointed analysis will not meet with universal approval. I had the chance to alter the manuscript but this would then have been less incisive and ultimately less useful. I see the human world at the start of the twenty-first century dominated by national self interest. We can wish that this were not so, but it is better to accept this reality and build a framework that uses self-determination to achieve positive outcomes.

I put forward a real-world framework. My analysis is based on logic, not ideology or vested interest. Some deductions are uncomfortable to read but we need to bring uncomfortable truths to the surface where they can be discussed and addressed.

The countries of the developed world have shown how to build modern societies. Mistakes have been made. Some of the flaws have been exposed, such as overreliance on fossil fuels. Other flaws are only now becoming apparent, such as squandering a whole range of non-renewable resources and sticking to economic metrics that do not align with human welfare. Those of us who live in the developed world need to understand, and accept, that the way we live is not perfect. We need to rein back from excessive self-belief that we manage model societies in order to be able make the required transition.

The developing world faces different challenges and has important choices to make. It is not for me or others in the West to insist how developing countries run their affairs but I suggest that they adopt greater self belief to run their societies in ways that suit their circumstances and their culture. Outside 'experts' should be treated with caution until the paradigm of sustainability is firmly established as the ruling logic.

My policy framework could be the unifying framework within which each country uses the power of self-determination to make progress despite the weakness of global governance and the intransigence of some nations. The current debate over the future for world society and the economy is important but more important is to move beyond the debate to action.

First, my assertion that economic globalization is no longer the paradigm to follow has to be accepted. World society is locked into a system that is entering a spiral of decline as the world comes up against the limits to growth. Backing off from the policies of globalization is the first step in allowing the development of a new world order.

Second, countries that understand the imperative to act should do so without delay, acting alone or in concert with regional groupings such as the EU. Appropriate action should be forced through in advance of reform of organizations such as the WTO, IMF and World Bank. This is how to drive reform from the bottom up on a country-by-country basis. The prize is to bounce forward to a system in which increased variety and fewer interdependencies brings greater macro stability (Chapter 11).

Third, the growing concern over the impacts of climate change in the more affluent countries should be channeled into making substantive progress in the transformation to sustainable societies. This will also require action to ensure the rapid diffusion of sustainable processes and green technologies to poorer countries (Chapter 10). Developing countries are more likely to be willing to follow a Western lead when they observe both tough action in developed societies and an openness to share the required solutions.

Economic globalization required decades to take firm hold because the economic theory that underpins it conflicts with a country's intuitive instinct to run its own affairs. Dismantling the edifice of globalization could happen almost overnight. Collapse of the world economy in combination with global food shortages (both of which look increasingly likely) could be the trigger. Alternatively we could move beyond globalization through deliberate choice to set up the circumstances of macro stability and security of supply of key resources.

The countries with the resources and most able to lead the transition should move first to set an example – and protect their future. I would like to see the rich countries use their power and influence to drive change. However it might be poorer countries with good natural resources that lead by choosing to manage their affairs in ways that suit their society and their people, showing the West how a sustainable society is possible.

For business, there are important choices as the world enters a period of disruption. Business will have to go with the flow of the changes demanded by society. When business leaders observe that society is reaching a tipping point, they will transform their corporations to match the new world order (Chapter 13).

The calls for action to take pressure off the ecosystem grow louder but, so far, have had little real effect. A way is needed to make real progress in the real world. This book shows a way to move quickly from talk to action.

References

Ackerman, F. 2008. The economics of collapsing markets. *Real-World Economics Review*, 48, 290. Available at: http:www.paecon.net/PAEReview/issue48/Ackerman48.pdf [accessed: 2 September 2009].

Adams, P. 1991. *Odious Debts: Loose Lending, Corruption, and the Third World's Environmental Legacy*. London: Earthscan.

Ames, G.J. 2008. *The Globe Encompassed: The Age of European Discovery, 1500–1700*. Harlow (UK): Prentice Hall, 102–3.

Anderson, V. 2006. Turning economics inside out. *International Journal of Green Economics*, 1 (1/2), 11-22.

ARRF 2009. *The Acid Rain Retirement Fund.* [Online]. Available at: http://www.usm.maine.edu/pos/arrf.htm#mission [accessed: 27 July 2009].

Bahn-Walkowiak, B., Bleischwitz, R., Bringezu, S., Bunse,M., Herrndorf, M., Irrek, W., Kuhndt, M., Lemken, T., Liedtke, C. and Machiba, T. 2008. *Resource Efficiency: Japan and Europe at the Forefront.* Germany: Federal Environment Agency. Available at: http://www.worldresourcesforum.org/files/file/RessEfficiency%20-%20Japan%20and%20Europe%20at%20the%20Forefront.pdf [accessed: 22 September 2009].

Bär, S. and Kraemer, R.A. 1998. European environmental policy after Amsterdam. *Journal of Environmental Law*, 10(2), 315–30.

Barbier, E.B. (2009). *Rethinking the Economic Recovery: A Global Green New Deal.* Geneva: UNEP.

Barry, J. 2007. Towards a model of green political economy: from ecological modernisation to economic security. *International Journal of Green Economics*, 1 (3/4), 446-462.

BerkShares 2009. *Local Currency for the Berkshire Region.* [Online]. Available at: http://www.berkshares.org [accessed: 27 August 2009].

Boyle, D. and Simms, A. 2009. *The New Economics: A Bigger Picture*. London: Earthscan.

Bremmer, I. 2009. State capitalism and the crisis. *McKinsey Quarterly*, July 2009. Available at: http://www.mckinseyquarterly.com/Strategy/Globalization/State_capitalism_and_the_crisis_2403 [accessed: 1 September 2009].

Burtraw, D. 2000. *Innovation under the Tradable Sulphur Dioxide Emission Permits Program in the US Electricity Sector*. Washington: Resources for the Future, 3. Available at: http://www.rff.org/documents/RFF-DP-00-38.pdf [accessed: 2 September 2009].

CarFree City USA 2009. *How Would You Prefer to Live?* [Online]. Available at: http://www.carfreecity.us [accessed: 1 September 2009].

Cato, M. 2009. *Green Economics: An Introduction to Theory, Policy and Practice.* London: Earthscan.

Cavanagh, J. and Mander, J. 2004. *Alternatives to Economic Globalization: A Better World Is Possible,* 2nd Edition. San Francisco: Berrett-Koehler.

Council of the European Union. 2006. *Review of the EU Sustainable Development Strategy (EU SDS) – Renewed Strategy,* 3–4. [Online]. Available at: http://ec.europa.eu/sustainable/docs/renewed_eu_sds_en.pdf [accessed: 2 September 2009].

Crawford, J.H. 2000. *Carfree Cities*. Netherlands: International Books.

Daly, H.E. 1991a. *Steady-State Economics: Second Edition with New Essays.* Washington DC: Island Press.

Daly, H.E. 1991b. Elements of environmental macroeconomics, in *Ecological Economics: The Science and Management of Sustainability,* edited by R. Costanza. New York: Columbia, 44–5.

Department for Environment, Food and Rural Affairs (DEFRA) 2008. *The Milk Road Map 2008*. London: HMSO, 12. Available at: http://www.defra.gov.uk/environment/business/pdf/milk-roadmap.pdf [accessed: 28 July 2009].

Design for Homes 2007. *Recommendations for Living at Superdensity*. London: Design for Homes. Available at: http://www.designforhomes.org/pdfs/Superdensity.pdf [accessed: 4 August 2009].

Diener, E. and Biswas-Diener, R. 2008. *Rethinking Happiness: The Science of Psychological Wealth*. Malden, MA: Blackwell Publishing.

Dresner, S. 2008. *The Principles of Sustainability,* 2nd Edition. London: Earthscan.

Ehrlich, P.R. 1968. *The Population Bomb*. New York: Ballantine Books.

Elliott, L., Hines, C., Juniper, T. et al. 2008. *A Green New Deal,* London: New Economics Foundation.

EPA 2009a. *Acid Rain and Related Programs 2007 Progress Report.* Washington: US Environmental Protection Agency, 1. Available at: http://www.epa.gov/airmarket/progress/docs/2007ARPReport.pdf [accessed: 2 September 2009].

EPA 2009b. *Understanding the Clean Air Act.* [Online: US Environmental Protection Agency]. Available at: http://www.epa.gov/air/caa/peg/understand.html [accessed: 23 July 2009].

Epstein, M.J. 2008. *Making Sustainability Work*. Sheffield, UK: Greenleaf Publishing.

Esty, D.C., Levy, M.A., Kim, C., de Sherbinin, A., Srebotnjak, T. and Mara, V. 2008. *2008 Environmental Performance Index.* New Haven: Yale Center for Environmental Law and Policy. Available at: http://epi.yale.edu [accessed: 2 September 2009].

EU 2008a. Consolidated version of the treaty on European Union, *Official Journal of the European Union* C 115/13. [Online]. Available at: http://eur-lex.europa.eu/LexUriServ/LexUriServ.do?uri=OJ:C:2008:115:0013:0045:EN:PDF [accessed: 2 September 2009].

EU 2008b. Emissions trading: Commission welcomes EP vote on including aviation in EU ETS. EU Press release 8 July 2008 [Online]. Available at: http://europa.eu/rapid/pressReleasesAction.do?reference=IP/08/1114 [accessed: 26 August 2009].

European Environment Agency 2010. *Natural resource accounting*. [Online]. Available at: http://glossary.eea.europa.eu/terminology/concept_html?term=natural resource accounting [accessed: 13 January 2010].

Friedman, T.L. 2005. *The World Is Flat: A Brief History of the Globalised World in the Twenty-first Century*. London: Allen Lane.

George, H. 1879. *Progress and Poverty: An Inquiry into the Cause of Industrial Depressions and of Increase of Want with Increase of Wealth: The Remedy*, 1912 Edition. New York: Doubleday, Page & Co. Available at: http://www.econlib.org/library/YPDBooks/George/grgPP.html [accessed: 2 September 2009].

Glennie, J. 2008. *The Trouble with Aid: Why Less Could Mean More for Africa*. London: Zed Books.

Global Footprint Network 2008. *Global Footprint Network's 2008 Edition National Accounts* [Online]. Available at: http://www.footprintnetwork.org/en/index.php/GFN/page/ecological_footprint_atlas_2008/ [accessed: 23 April 2009].

Gore, A. 2000. *Earth in the Balance: Ecology and the Human Spirit*. Boston: Houghton and Mifflin.

Gore, A. 2006. *An Inconvenient Truth: The Planetary Emergency of Global Warming and What We Can Do About It*. New York: Bloomsbury.

Greenwald, B. and Stiglitz, J. 2006. *A Modest Proposal for International Monetary Reform*. Paper to the American Economic Association, Boston, 4 January 2006. Available at: http://www.ofce.sciences-po.fr/pdf/documents/international_monetary_reform.pdf [accessed: 29 July 2009].

Hails, C. (ed.) 2008. *Living Planet Report 2008*. Switzerland: WWF, 1–2.

Hardin, G. 1968. The tragedy of the commons. *Science*, 162(3859), 1243–8.

Hargroves, K.J. and Smith, M.H. (eds) 2005. *The Natural Advantage of Nations: Business Opportunities, Innovation and Governance in the 21st Century*. London: Earthscan.

Harris, M.S. and Fraser, I. 2002. Natural resource accounting in theory and practice: a critical assessment. *The Australian Journal of Agricultural and Resource Economics*, 46(2), 139–192. **Abstract is a**vailable at http://ssrn.com/abstract=316014 [accessed: 27 July 2009].

Hart, S.L. 2005. *Capitalism at the Crossroads: The Unlimited Business Opportunities in Solving the World's Most Difficult Problems*. New Jersey: Wharton School Publishing, 3–54.

Hau, H. and Rey, H. 2006. Exchange rates, equity prices, and capital flows. *The Review of Financial Studies*, 19(1), 273–317.

Hawken, P., Lovins, A.B. and Lovins, L.H. 1999. *Natural Capitalism: The Next Industrial Revolution*. London: Earthscan.

Heinemann, V. 2007. Current developments in international trade – an opportunity for a new progressive approach in economic policies. *International Journal of Green Economics*, 1 (3/4), 351-373.

Hines, C. 2000. *Localization: A Global Manifesto*. London: Earthscan.

Hong, B.D. and Slatick, E.R. 1994. Carbon dioxide emission factors for coal. *Quarterly Coal Report*, 94(1). Washington: Energy Information Administration, 1–8. Available at: http://www.eia.doe.gov/cneaf/coal/quarterly/co2_article/co2.html [accessed: 23 July 2009].

Hopkins, R. 2008. *UK: The Transition Handbook: From oil dependency to local resilience.* UK: Green Books.

IIED 2007. *Kenya: Africa Caught in the Tough 'Food Miles' War with UK*. [Online: International Institute for Environment and Development, 7 November 2007] Available at: http://www.agrifoodstandards.net/en/news/global/kenya_africa_caught_in_the_tough_food_miles_war_with_uk.html [accessed: 4 August 2009].

IMF 2008. *World Economic Outlook October 2008*, 259. Available at: http://www.imf.org/external/pubs/ft/weo/2008/02/pdf/text.pdf [accessed: 25 August 2009].

Inglehart, R., Foa, R., Peterson, C. and Welzel, C. 2008. Development, freedom, and rising happiness: a global perspective (1981–2007). *Perspectives on Psychological Science*, 3(4), 264–85.

IPCC 2007. *Climate Change 2007: Synthesis Report. Contribution of Working Groups I, II and III to the Fourth Assessment Report of the Intergovernmental Panel on Climate Change*. Geneva: IPCC. Available at: http://www.ipcc.ch/pdf/assessment-report/ar4/syr/ar4_syr.pdf [accessed: 24 July 2009].

Jackson, T. 2009. *Prosperity without Growth? The Transition to a Sustainable Economy.* London: Sustainable Development Commission (SDC), 79–81.

Jacques, M. 2009. *When China Rules the World: The Rise of the Middle Kingdom and the End of the Western World*. London: Allen Lane.

Jayachandran, S. and Kremer, M. 2006. Odious debt. *The American Economic Review*, 96(1), 82–92.

Kennet, M. and Heinemann, V. 2006. Green Economics: setting the scene. Aims, context, and philosophical underpinning of the distinctive new solutions offered by Green Economics. *International Journal of Green Economics*, 1 (1/2), 68-102.

Ki-moon, B. 2009. *The Global Compact: Creating Sustainable Markets*. Speech at the World Economic Forum, Davos, Switzerland, 29 January 2009. Available at: http://www.unep.org/Documents.Multilingual/Default.asp?DocumentID=560&ArticleID=6059&l=en&t=long [accessed: 2 September 2009].

King, D.A. 2004. Climate change science: adapt, mitigate, or ignore? *Science*, 303(5655), 176–7.

Kitzes, J., Buchan, S., Galli, A. et al. 2008. *Report on Ecological Footprint in China*. Beijing: WWF China, 13. Available at: http://www.footprintnetwork.org/en/index.php/GFN/page/publications/ [accessed: 29 July 2009].

Knox, P. 2005. Creating ordinary places: Slow cities in a fast world. *Journal of Urban Design*, 10(1), 1–11.

Kochan, B. 2007. Squeezing suburbia. *Planning*, 28 September 2007. Available at: http://www.planningresource.co.uk/careers/features/740770/Squeezing-suburbia [accessed: 4 August 2009].

Krugman, P. and Obstfeld, M. 2009. *International Economics: Theory and Policy*, 8th Edition. Boston: Pearson Education, 4.

Kunstler, J.H. 1996. *Home from Nowhere: Remaking Our Everyday World for the 21st Century*. New York: Simon & Schuster.

Kunstler, J.H. 2005. *The Long Emergency: Surviving the Converging Catastrophes of the Twenty-first Century*. New York: Grove/Atlantic.

Kuznets, S. 1934. *National Income, 1929–1932*. 73rd US Congress, Senate document no. 124, 7. Available at: http://library.bea.gov/u?/SOD,888 [accessed: 4 August 2009].

Kuznets, S. 1941. *National Income and Its Composition, 1919–1938*. New York: National Bureau of Economic Research Publications, No. 40.

Lawn, P. 2005. Is a democratic–capitalist system compatible with a low-growth or steady-state economy? *Socio-Economic Review*, 3(2), 209–32.

Layard, R. 2005. *Happiness: Lessons from a New Science*. London: Allen Lane.

LETSlink UK 2009. *So What are LETS?* [Online]. Available at: http://www.letslinkuk.org [accessed: 27 August 2009].

Lietaer, B. 1997. *From the Real Economy to the Speculative*. Remarks by Bernard Lietaer at International Forum on Globalization (IFG) seminar, 15 December 1997 [Online]. Available at: http://www.hartford-hwp.com/archives/25/062.html [accessed: 29 July 2009].

Lunn, C. 2006. The role of green economics in achieving realistic policies and programmes for sustainability. *International Journal of Green Economics*, 1(1/2), 37-49.

Lüthi, D., Le Floch, M., Bereiter, B. et al. 2008. High-resolution carbon dioxide concentration record 650,000–800,000 years before present. *Nature*, 453, 379–82.

Machiavelli, N. 1515. *The Prince*, translated by Marriott, W. K. 1908, Chapter III. Available at: http://www.constitution.org/mac/prince03.htm [accessed: 2 September 2009].

Marks N., Abdallah S., Simms, A. and Thompson S. 2006. *The Happy Planet Index*. London: New Economics Foundation (nef).

McDonough, W. and Braungart, M. 2002. *Cradle to Cradle: Remaking the Way We Make Things*. New York: North Point Press.

McGranahan, G., Balk, D. and Anderson, B. 2007. The rising tide: assessing the risks of climate change and human settlements in low elevation coastal zones. *Environment and Urbanization* 19(1), 17–37.

McManners, P.J. 2007. *Cities for People: Removing Cars from Urban Life*. Paper presented at the World Institute for Development Economics Research of the United Nations University (UNU-WIDER) project workshop, Beyond the Tipping Point: Development in an Urban World, London School of Economics and Political Science, 19–20 October 2007.

McManners, P.J. 2008. *Adapt and Thrive: The Sustainable Revolution*. UK: Susta Press.

McManners, P.J. 2009. *Victim of Success: Civilization at Risk*. UK: Susta Press.

MDG Monitor 2009. *Ensure Environmental Sustainability* [Online]. Available at: http://www.mdgmonitor.org/goal7.cfm [accessed: 25 August 2009].

Meadows, D.H., Meadows, D.L. and Randers, J. 1972. *The Limits to Growth*. New York: Universe Books.

Meadows, D.H., Randers, J. and Meadows, D. 2004. *Limits to Growth: The 30-Year Update*. Vermont: Chelsea Green Publishing Company.

Millennium Ecosystem Assessment 2005. *Ecosystems and Human Well-being: Synthesis*. Washington DC: Island Press.

Moran, A. 2006. The public transport myth. *IPA Review*, October 2006, 8–11. Available at: http://www.ipa.org.au/library/58-3%20MORAN.pdf [accessed: 2 September 2009].

Netherlands Environment Assessment Agency 2008. *Global CO_2 Emissions: Increase Continued in 2007* [Online 13 June 2008]. Available at: http://www.pbl.nl/en/publications/2008/GlobalCO2emissionsthrough2007.html [accessed: 27 July 2009].

Newman, P., Beatley, T. and Boyer, H. 2009. *Resilient Cities: Responding to Peak Oil and Climate Change*. USA: Island Press.

OECD 2009. *Agricultural Policies in OECD Countries Monitoring and Evaluation 2009*. Paris: OECD. Available at: http://www.oecd.org/dataoecd/37/16/43239979.pdf [accessed: 22 September 2009].

OPT 2009. *YouGov Survey Results*. [Online.] Available at: www.optimumpopulation.org/submissions/YouGov11Jul09.xls [accessed: 27 August 2009].

Ostrom, E. 2009. *Sveriges Riksbank Prize Lecture, 8 Dec. 2009*. [Online.] Available at: http://nobelprize.org/nobel_prizes/economics/laureates/2009/

Porritt, J. 1994. *Seeing Green: The Politics of Ecology Explained*. Oxford: Blackwell, 181.

Porritt, J. 2005. *Capitalism as if the World Mattered*. London: Earthscan, 3–100.

Reardon, J. 2007. Comments on 'Green Economics: setting the scene. Aims, context, and philosophical underpinning of the distinctive new solutions offered by Green Economics'. *International Journal of Green Economics*, 1(3/4), 532-538.

Ricklefs, M.C. 1991. *A History of Modern Indonesia Since c.1300*, 2nd Edition. London: MacMillan, 110.

SDC 2009a. *Sustainable Development Commission* [Online]. Available at: http://www.sd-commission.org.uk/pages/about-us.html [accessed: 31 August 2009].

SDC 2009b. *Our Principles* [Online]. Available at: http://www.sd-commission.org.uk/pages/our-principles.html [accessed: 31 August 2009].

SDC 2009c. *Breakthroughs for the Twenty-First Century*. London: SDC.

Smith, A. 1759. *The Theory of the Moral Sentiments*, Part VI, Chapter III. Available at: http://www.adamsmith.org/smith/tms-intro.htm [accessed: 25 July 2009].

Smith, A. 1776. *An Inquiry into the Nature and Causes of the Wealth of Nations*, Part IV, Chapter II. Available at: http://www.adamsmith.org/smith/won-intro.htm [accessed: 25 July 2009].

Smith, A., Watkiss P., Tweddle, G. et al. 2005. *The Validity of Food Miles as an Indicator of Sustainable Development*. UK: DEFRA, ii.

Stewart, S.I., Hammer, R.B., Radeloff, V.C., Dwyer, J.F. and Voss, P.R. 2003. *Mapping Housing Density across the North Central U.S., 1940–2000* [Online slide show]. Available at: http://www.ncrs.fs.fed.us/IntegratedPrograms/lc/pop/hd/title.htm [accessed: 2 September 2009].

Stiglitz, J.E. 2006. *Making Globalization Work*. London: Allen Lane.

Tainter, J. 1990. *The Collapse of Complex Societies*. Cambridge: Cambridge University Press.

Tans, P. 2009. *Trends in Atmospheric Carbon Dioxide – Mauna Loa*. [Online]. Available at: http://www.esrl.noaa.gov/gmd/ccgg/trends/ [accessed: 24 July 2009].

The Economist 2009. Enter the dragon. *The Economist*, 11 July 2009, 80–81.

The Royal Society 2006. *Royal Society Submission to the Environmental Audit Committee's Inquiry into the UN Millennium Ecosystem Assessment*. RS Policy Document 30/06. Available at: http://royalsociety.org/displaypagedoc.asp?id=24389 [accessed: 24 July 2009].

The World Commission on Environment and Development 1987. *Our Common Future*. UK: Oxford University Press.

Tinsley, S. and George, H. 2006. *Ecological Footprint of the Findhorn Foundation and Community*. Scotland: Sustainable Development Research Centre, 4. Available at: http://www.ecovillagefindhorn.org/docs/FF%20Footprint.pdf [accessed: 29 July 2009].

Tobin, J. 1978. A proposal for international monetary reform. *Eastern Economic Journal*, 4(3–4), 153–9.

Turner, G. 2008. A comparison of limits to growth with thirty years of reality. *Commonwealth Scientific and Industrial Research Organisation (CSIRO) Working Paper Series*, June 2008. Collingwood, Australia: CSIRO Publishing, 37. Available at: http://www.csiro.au/files/files/plje.pdf [accessed: 24 July 2009].

Ullah, F., Shields, A. and Crees, J. 2009. *Sustainable Development in Government: Challenges for Government 2008*. UK: Sustainable Development Commission.

UN 1992. *Report of the United Nations Conference on Environment and Development* (Rio de Janeiro, 3–14 June 1992) [Online]. Available at: http://www.un.org/documents/ga/conf151/aconf15126-1annex1.htm [accessed: 24 July 2009].

UN 2000a. *United Nations Millennium Declaration*. General Assembly Resolution 55/2, 18 September 2000.

UN 2000b. *Millennium Report of the Secretary-General of the United Nations*. New York: United Nations, 56. Available at: http://secint24.un.org/millennium/sg/report/ch4.pdf [accessed: 3 August 2009].

UN 2001. *Road Map towards the Implementation of the United Nations Millennium Declaration*. General Assembly Resolution 56/326, 6 September 2001, 56–8.

UN 2008. *The Millennium Development Goals Report 2008*. New York: United Nations, 3.

UN 2009. *World Population Prospects: The 2008 Revision, Highlights*. Working Paper No. ESA/P/WP.210, vii. Available at: http://www.un.org/esa/population/publications/wpp2008/wpp2008_highlights.pdf [accessed: 29 July 2009].

UNCTAD 2008. *Trade and Development Report 2008*. Geneva: UNCTAD, iv. Available at: http://www.unctad.org/en/docs/tdr2008fas_en.pdf [accessed: 28 July 2009].

UNDP 1990. *Human Development Report 1990*. New York: Oxford University Press.

UNEP 2005. *Environment and Trade – A Handbook*. Geneva: UNEP, 29.

UNEP 2009. *UNEP Organization Profile* [Online]. Available at: http://www.unep.org/PDF/UNEPOrganizationProfile.pdf [accessed: 4 August 2009].

US Congress 2009. *American Clean Energy and Security Act of 2009*. Bill H.R. 2454, 111th Congress, sponsored by Henry Waxman and Edward Markey, Washington: US Congress.

Vanhanen, H., Toppinen, A., Tikkanen, I. and Mery, G. 2007. *EFI Policy Brief 1: Making European Forests Work for People and Nature*. Joensuu, Finland: European Forestry Institute, 4. Available at: http://www.efi.int/files/attachments/publications/efi_policy_brief1_net.pdf [accessed: 24 July 2009].

Vickers, T. 2007. *Location Matters: Recycling Britain's Wealth*. London: Shepheard-Walwyn.

Victor, P. 2008. *Managing Without Growth: Slower by Design, Not Disaster*. Cheltenham, UK: Edward Elgar.

WBCSD 2001. *World Mobility at the End of the Twentieth Century and its Sustainability*. Geneva: World Business Council for Sustainable Development. Available at: http://www.wbcsd.org/web/projects/mobility/english_full_report.pdf [accessed: 4 August 2009].

WCED 1987. *Our Common Future*. Oxford: Oxford University Press, 43.

Weizsäcker, E., Lovins, A.B. and Lovins, L.H. 1998. *Factor Four: Doubling Wealth – Halving Resource Use*. London: Earthscan.

Williamson, J. 1990. What Washington Means by Policy Reform, in *Latin American Adjustment: How Much Has Happened?*, edited by J. Williamson. Washington, D.C.:Institute for International Economics, 5-20.

Woodin, M. and Lucas, C. 2004. Green Alternatives to Globalization: A Manifesto. London: Pluto Press.

World Bank 2009a. *Measuring Poverty at the Country Level* [Online]. Available at: http://go.worldbank.org/K7LWQUT9L0 [accessed: 25 August 2009].

World Bank 2009b. *2009 World Development Indicators*. Washington, DC: The World Bank.

WTO 1994. *Decision on Trade and Environment*. Results of the Uruguay Round of Multilateral Trade Negotiations, Marrakesh, 15 April 1994. [Online]. Available at: http://www.wto.org/english/docs_e/legal_e/56-dtenv_e.htm [accessed: 28 July 2009].

WTO 2009a. *World Total Merchandise Exports in US dollars at Current Prices*. [Online: WTO Statistics database]. Available at: http://stat.wto.org [accessed: 2 September 2009].

WTO 2009b. *Items on the CTE's Work Programme*. [Online]. Available at: http://www.wto.org/english/tratop_e/envir_e/cte00_e.htm [accessed: 29 July 2009].

Index

Index

Acid Rain Retirement Fund (ARRF) 63
Ackerman, F. 199
Adams, Patricia 127
Afghanistan 106
Africa 18, 27, 35–6, 53, 132, 184
Agenda 21. 23
agrarian economy 105, 201
agriculture 29, 38, 87, 89–90, 192
  exports 68, 105
aid 35, 137, 155
  agencies 36
airline industry 170
aluminium smelting 94
Amazon 27
Ames, G.J. 174
Anderson, B. 4
Anderson, V. 55
Antarctic 5
anthropogenic forces 31
Arab nations 11
Arctic 5
Asia 36, 119,
  crisis 122
Association of Southeast Asian Nations (ASEAN) 51, 122
Australia 11, 77
automation 57, 102–103
aviation 64
air freight 68
Bahn-Walkowiak, B. 86
balance 9, 11–12, 37, 108, 157, 198
Balk, D. 4
bancor 121
Bangladesh 36, 77, 79, 104, 106
banking 113–117, 130
  trust 114
  world central bank 122
Bär, S. 51
Barbier, E.B. 200
Barry, J. 21
bartering 112–113, 123–24
Beatley, T. 189
Beddington Zero Energy Development (BedZED) 108
Bereiter, B. 4
BerkShares 124
Bhutan 48, 53, 80
biodiversity 42
biofuel 27, 29, 68, 92, 94, 178
biosphere xvii, 32
black economy, *see* economy
Bleischwitz, R. 86
Bolivia 176–7
borders 88
Bosnia 50
Boyer, H. 189
Boyle, D. 55
Braungart, Michael 66, 94
Brazil 104–5
Bremmer, I. 176, 179

Bretton Woods 121
Bringezu, S. 86
British Empire 32
Brown, Gordon 114
Brundtland Report 21
Buchan, S. 105
Bunse,M. 86
Burtraw, D. 62
business 168, 170
 as agent for change 168, 171–2, 206–7
business schools 158, 168, 175, 179, 204, 207

Cameroon 184
Canada 50
cap-and-trade 62
capital 111–130
 expenditure 103
 flows 99, 118, 125
 markets 111
capitalism 6, 26, 27, 43, 76, 181, 201
 laissez-faire 39–40, 177
carbon trading 55–6, 64, 92, 169, 206
 auction 63
 limitations of 64–5
 national carbon markets 65–6
 price of carbon 63, 199
Carbon dioxide
 emissions 37, 63–4, 78, 92, 160
 levels in the atmosphere 4
 release of 5
CarFree City USA 189
Carnegie, Andrew 80
cars 188
Cato, M. 55
Cavanagh, J. 147
cement manufacturing 64
Charter for Human Rights 37
China 77–79, 94, 104–106, 119, 175, 201
Chu, Steven 162
cities 41
 and cars 41, 49, 168, 183, 187–91
 design 160, 183–9
 megacities 41, 188
 urban villages 42
civilization xvii, 32, 39, 48, 131–2, 208
 collapse of 19, 32, 81, 195
 future of 22
Clean Air Act (US) 62
Clean Development Mechanism (CDM) 65
climate change 4, 5, 15, 19, 29, 42, 88, 155, 162, 192, 203, 205
 refugees 33
 tipping point 5
Clinton, Bill 15
Club of Rome 22, 196–7
coal 62, 64–5, 88, 92–4
Collateralized Debt Obligation (CDO) 116
commodity flows 13, 17, 83–97, 197
commons, 8, 35
 tragedy of the 8, 77
commitment 10–11
community 11, 16, 42, 77, 80, 143
 engagement 183–4
 human-scale 181–93
 local 11,18, 40, 99
 sense of 11, 107, 158
 world 11, 33
comparative advantage 40, 134–5, 176
computers 17, 95, 102, 131–4, 204
 chips 133, 136
consumption 18, 37, 78, 87, 100, 104–105, 107
copyright 134; *see also* intellectual property rights
corporate social responsibility (CSR) 169
corporate strategy, *see* strategy

corporations 110, 136–7, 169–179; *see also* global corporations
corruption 32, 53, 88, 107, 154, 174, 195
Council of the European Union 165
cradle to cradle 66, 94, 103; *see also* lifecycle
Crawford, J.H. 188
Crees, J. 164
Croatia 50
Cuba 158–9
culture 47, 48, 107–108, 133, 184, 189
currencies 112–3, 120–23, 150
  local 123–4, 182; *see also* Local Exchange Trading Schemes (LETS)
  national 123

Daly, Herman 23, 30
democracy 144, 160, 206
Department for Environment, Food and Rural Affairs (DEFRA) 91, 163
Department of Energy and Climate Change (DECC) 160
Design for Homes 187
developed countries 21, 33, 65, 70, 89, 101–2, 125, 134, 190
developing countries 22–3, 36, 63, 65, 78, 102, 139,146, 190
Diener, Ed 18
distance-to-market tax 66–7, 90
division of labour 40
dolphin protection 38
Dresner, Simon 30
drugs 136
Dutch East India Company 174
Dwyer, J.F. 187

Earth 27, 36, 43, 95, 135, 195, 208
  capacity/resources 30, 77–8, 97, 100, 202
Earth Summit 23, 148
ecological capacity 36, 37, 38, 77, 104–105; *see also* Earth
  debtors 104, 105
  footprint 77
economic development 6, 8
economic globalization, *see* globalization
economics 10, 31, 181, 190, 197, 204
  green, 35–6, 55–71
  interdependence 43
  market 13, 60, 198
  methods 10
  policy 12, 196
  theory 40
economists xvii, 22, 97, 116, 125, 167–8, 196, 200, 202–3
  green 21, 96
economy, the xvi, xviii–xix, 9, 13–4, 16, 25, 28–9, 43, 51, 112, 196
  black 124–5
  green xv, 35–6, 207
  real 112,117, 128, 130
ecosystem 19, 27, 30, 32, 42–3, 70, 97, 134, 199, 203, 205
efficiency gains 17, 27, 59
Ehrlich, Paul 104
Elkington, John 21
Elliott, L. 200
emigration 107–108
emissions trading 62–6; *see* also EU ETS
employment 16, 103, 123, 200; *see also* labour
empowerment 36, 50
energy 63–5, 83, 92–31, 160, 162, 175, 177–8; *see also* renewable energy

energy security 206
environment xv–xix, 3, 6, 9–10, 12–4, 17–8, 21, 25–6, 28–31, 35, 39–41, 75, 78, 87–8, 108, 143–5, 148–9
Environmental Performance Index 165
Environmental Protection Agency (EPA) 62, 63
environmental policy 38, 165, 199
  protection 9,10, 85, 87, 166–7, 173, 196
  regulations 38, 178, 155, 198
  taxes 57; *see also* taxation
environmentalists xviii, 104
Enron 174
Epstein, Marc 169
equity 25, 166, 199
equity markets 128
Esty, D.C. 165
Ethiopia 68
euro, the 119–122, 125
  euro zone 120
Europe 11, 27, 41, 51, 78, 86, 105, 120, 136, 174, 199, 201
European Central Bank 121
European commission 63, 92, 165–7
European Environment Agency 69
European Parliament 165–6
European Union xvi, 11, 51
  treaty 35
European Union Emission Trading System (EU ETS) 56, 63, 64, 199
externalities 40, 196–7
extractive industries 87

family 9, 11–2, 39–40, 47, 99, 101, 107, 110
farming 69
federal structure 50, 51, 63
finance 111–130; *see also* global finance
financial crash 10
  (1929) 123
  (2008) 43, 111, 179, 197, 199
    subprime trigger 115–116
financial services 115, 138
Findhorn Foundation and Community 108
Finland 128, 165
Foa, R. 36
food xvi, 6, 29, 38, 41, 66–8, 83, 89–94, 109–10, 182, 192
  miles 67
foreign direct investment 172
forestry 87–8
fossil fuel 37, 64, 88, 92, 170–1, 179, 206
  taxation 59, 160, 205
France 193
Fraser, Iain 80
free trade 17, 51, 84, 89, 95, 99, 106, 132
French, H. 200
Friedman, Thomas 171
fuel poverty 60

Galli, A. 105
Gardner, G. 200
gas 62, 64–5, 88, 92, 176
Gates, Bill 81
GATT 38
GDP 4, 17, 18, 80, 133, 197, 203
General Motors (GM) 179
genocide 110
George, Henry 58
George, Heather 108
Germany 50, 120, 128
Glennie, J. 154
global corporations 175–6

global economy, *see* world economy
global finance 111–112
Global Footprint Network 77, 105–6, 108, 201
Global Green New Deal (GGND) 200
global hectare (gha) 77
globalization xvii, 3, 7, 12, 14–5, 17, 19, 33, 52, 75, 91, 101, 123, 128, 133–4, 169, 171, 179, 196, 202–3
gold standard 113
Gore, A. 15
government 44, 50, 103, 155, 157–68, 204, 207
  income 57, 60, 124
  intervention 55, 64, 115, 129
  ownership 117–118
  policy framework 172
governance 27, 110, 135, 163, 172
  environmental 7
  global 7, 16, 44–5, 61, 144, 146–7, 199
Grameen Bank 36
Great Depression 14, 113, 123, 200
green economics, *see* economics
green economists, *see* economists
green taxation, see taxation
green technology 137
Greenwald, B. 121
gross domestic product, *see* GDP
growth 16, 21–22, 167, 196
  export-led 138, 150, 153
  limits to 22, 25, 196
  low-growth 16–7, 22, 167

Hails, C. 36, 77, 104, 108
Hammer, R.B. 187
happiness 18, 36, 80, 133, 205
  Gross National Happiness (GNH) 48, 80
  The Happy Planet Index (HPI) 158
Hardin, Garrett 8, 77
Hargroves, K.J. 169
Harris, Michael 80
Hart, Stuart 6, 104
Hau, H. 112
Hawken, P. 169
Heavily Indebted Poor Countries (HIPC) 126
Heinemann, Volker 21, 55, 96
Herbolzheimer, Emilio xix
Herrndorf, M. 86
Hines, Colin 12, 200
Holdren, John 162
Hong, B.D. 65
Hopkins, Rob 181
housing density 187
Human Development Index (HDI) 80
human progress xvii, 18, 31, 39, 202–3
human-scale 42, 181, 185, 193
hydrogen 93, 178

Iceland 89, 94
IMF 4, 14, 38, 43, 121–2, 125–6, 158, 196, 202
immigration 99, 101–102, 107–9
income 18, 57
  tax 58–9
India 78, 104–5
industrial processes 90, 96
Industrial Revolution 5, 14, 26, 192, 195
industrialization 19, 139
industry 149, 170–71, 176
inequity 25
Information Technology 103, 130–31
infrastructure 7, 13, 19, 28, 59, 102, 160–61, 178, 189
Inglehart, R. 36
intellectual property rights (IPR) 95, 134, 137–8

Intergovernmental Panel on Climate Change, see IPPC
International Forum on Globalization (IFG) 147
International Institute for Environment and Development (IIED) 68
International Monetary Fund, *see* IMF
IPCC 4–5, 93
Iraq 127
Irrek, W. 86

Jackson, Tim 17
Jacques, Martin 201
Japan 86, 89, 104, 119
Jayachandran, S. 127
jobs 58, 101–3, 172, 184, 197, 200
Juniper, T. 200

Kennet, M. 21, 55
Kenya 68
Keynes, John Maynard 121
Ki-moon, Ban 25, 143, 169
Kim, C. 165
King, David 4
Kitzes, J. 105
knowledge 131
  aid 137–9
  economy 95, 131–9,176
Knox, P. 181
Kochan, B. 187
Korea 89
Kraemer, R.A. 51
Kremer, M. 127
Krugman, P. 96
Kuhndt, M. 86
Kunstler, James 189
Kuznets, Simon 203
Kyoto Protocol 65

labour 101–3
Land and landowners 26
land tax 58–9, 68, 124
land use 26–30, 68, 71
Land Rover 18
Lawn, P. 167
Layard, Richard 80
Le Floch, M. 4
Lemken, T. 86
LETSlink, UK 124
Levy, M.A. 165
Liedtke, C. 86
Lietaer, Bernard 112
lifecycle 39, 67, 168; *see also* cradle to cradle
lifestyle 26, 77, 79, 100, 132, 159, 182, 188, 204
Limits to Growth, the 22, 25, 196–7
liquid sunshine 93–4, 178
Lloyds TSB 118
Local Exchange Trading Schemes (LETS) 124; *see also* currencies, local
localization 12, 38, 41, 76
  of production 95
London School of Economics xvi, xix, 48, 189
Lovins, A.B. 169
Lovins, L.H. 169
low-carbon 65–66, 88, 161, 164, 200
low growth economy 17
Lubchenco, Jane 162
Lucas, Caroline 12
Lunn, C. 21
Lüthi, D. 4
Luxembourg 47

Macedonia 50
Machiavelli, Niccolò 195
Machiba, T. 86

Malawi 106
Maldives 79
Mander, J. 147
manufacturing 64, 66–7, 94–5, 133, 136–8
Mara, V. 165
market economics, *see* economics
Markets 38, 56, 61, 71, 198–9
  carbon, *see* carbon trading
  free/open xv–xvi, 17, 27, 40, 60, 77, 84, 99, 101, 120, 125, 132, 169, 171, 202
  oversight of 44, 56
Markey, Edward 56
Marks N., Abdallah S., Simms, A. and Thompson S.
McDonough, W. 66, 94
McGranahan, G. 4
McManners, Peter 7, 19, 43, 50, 66, 70, 81, 92, 94, 96, 103, 107, 128, 170, 178, 188, 191
MDG Monitor 24
Meadows, D.H. 22, 196
Meadows, D.L. 22, 196
measurement 18–9, 80
media age 161, 173
megacities, *see* cities
Mery, G. 27
Methane 5
Mexico 38
migration 99
military operations 35
milk supply chain 91
Millennium Development Goals 24
Millennium Ecosystem Assessment 6, 13, 14
mining 38, 155, 173
Mission Command 35
Morales, Evo 177
Moran, A. 186
multinational corporations (MNC) 129, 135, 137–8, 169, 171–2, 174
mutual ownership 117–118

nation state xvi, 11, 47, 49–51, 54, 143, 157, 198; *see also* state
nationalization 173
nationalism 13
national parks 27, 69–70, 192
nationality 47
natural capital 32
natural resource accounting 69
nature 26, 27, 29, 31, 70, 191–2
neo-conservatives xv
neo-liberals xv, 132
Netherlands 174
Netherlands Environment Assessment Agency 78
New Economics Foundation (nef) 55, 158, 200
New Green Deal 200
New Zealand 79, 104, 165
Newbury Building Society 117
Newman, P. 189
non-governmental organizations (NGO) 6, 207–8
North American Free Trade Agreement (NAFTA) 51
Northern Ireland 183
Northern Rock 116, 118
Norway 89, 165
nuclear power 93, 144
nuclear weapons 134, 143–4

Obama, Barack 157, 162
Obstfeld, M. 96
odious debt 127
OECD 89
Oil 19, 29, 38, 93
  industry 170–1, 176

reserves 42, 176
sands 171
Optimum Population Trust (OPT) 108
Ostrom, Elinor 35
Paleocene-Eocene Thermal Maximum 5
pariahs 170
Peterson, C. 36
Planet Earth, see Earth
politicians 14–15, 18, 57, 61, 64, 69, 90, 99, 125, 152, 157, 159–61, 166, 168, 189, 199, 206, 208
political leadership 205–6
Pollution 27, 32, 57, 60–63, 71, 95, 97, 149, 188, 192
markets 62
Population 29, 31, 76, 99–110
ageing 109
growth 42, 90, 99–101, 110, 155, 205
reduction 106–107
Porritt, Jonathan 6, 59, 164
Poverty 9, 24, 26, 36
Power 10
primacy of the state, *see* state
privatization 52, 53
production, *see* manufacturing
protectionism xix, 52
proximization 12, 13, 16, 17, 76, 79, 101, 123, 125, 133, 179, 196–8, 202, 208
public transport 60, 186
quality of life 18, 21, 187, 190–91, 193, 204–5

Radeloff, V.C. 187
rainforest 90
Randers, J. 22, 196
Reardon, Jack 55
recession 13, 52, 83; *see also* financial crash
recycling 39, 59, 66–7, 95–6, 103, 176
regulation 55–6, 61, 96, 114, 160, 177
renewable energy 67,92, 94, 160, 178, 188, 206
Renner, M. 200
research and development 135–6
reserve currency 119–21
Rey, H. 112
Ricklefs, M.C. 174
Rio Declaration 23
Rio de Janeiro 23
robots 95, 102, 136
Roman Empire 32
Russia 53, 104, 175

Sack, Alexander 127
sea level rise 4
Second World War 50, 134, 145–6, 203
security of supply 66, 85–7, 89, 97, 108
self-determination 48, 75–6
self-interest 33, 45, 75–6, 87, 145, 169, 173, 182, 199
selfish altruism 181, 184
Serbia 50
shareholder value 175, 207
Sherbinin de, A. 165
Shell oil company 171
Shields, A. 164
Simms, A. 55
Slatick, E.R. 65
Slovenia 50
slow cities movement 181
Smith, Adam 40, 205
Smith, Alison 67
Smith, James 171
Smith, M.H. 169
social provision 10, 29–30
social security 103

Society 9, 26–29, 33; *see also* world society
solar power 94, 178
Soros, George 130
sovereign state 10, 37
sovereign debt 121
spatial distribution 185–7
Special Drawing Right (SDR) 121
species xvii–xviii, 42–3, 184, 192, 208
speculation 59, 84, 112, 118, 127–9, 174
Srebotnjak, T. 165
state 47–54, 49, 50–1, 175
  primacy of the state 12–3, 54, 198
Stewart, S.I. 187
Stiglitz, Joseph 121, 125
strategy, corporate 136, 175, 178, 207
strategy for sustainable development 165–6
subsidiarity 12, 35–45, 36, 143, 198
suburbanization 41, 188, 190
sulphur dioxide emissions 62
supply chain 55, 90, 91, 171, 177–8
  localizing of 66–68
supply routes 108, 110
sustainability xviii, 3, 6–7, 9, 12, 15, 21–33, 36–7, 41, 75–6, 86–7, 102, 150–51, 155, 172, 197, 206–8
  definition 21–2
  model of 26, 28–30
  principles xviii–xix, 28, 30–32, 77–8, 68, 80, 152, 163, 169, 177
  theory 25
sustainable
  development 21, 25, 51, 148
  economy 64, 81, 163–4
  policy framework 76, 155, 157, 159, 161, 167, 208
  taxation 57–58
  trade, *see* World Sustainable Trade Organisation
Sustainable Development Commission (SDC) 157, 163–65, 167
Sustainable Revolution 7, 56, 165, 169, 179, 198
Sweden 79, 128, 165
Switzerland 50, 86, 89, 123, 165
synergy 17

Tainter, J. 32
Tans, P. 4
Tanzania 68
tariffs 85, 89
taxation 56, 173
  environmental 57
  for social outcomes 58
  green 56, 68, 124–125
  land, *see* land tax
  policy 55
  share transactions 128
technology 104, 117, 136
Thatcher, Margaret 53
The Economist 201
The Royal Society 6
Third World debt 126
throwaway society 39, 96
Tikkanen, I. 27
Tinsley, S. 108
Tito, Marshall 50
Tobin, James 128
Toppinen, A. 27
Toyota 179
Trade 17, 54, 95–6
Tragedy of the Commons, *see* commons
transition taxes 59–60
Transition Towns movement 181
transportation 66, 185–7
tribe 39, 107, 110

Turner, G. 22
Turner, Paul xix
Tweddle, Geoff 67

Ullah, F. 164
United Kingdom 53, 108–109, 113–4, 128, 160, 163
  EPI index 165
  UK government 91, 116, 129, 130, 160
  HPI index 158
United Nations 6, 10, 100, 128
  Secretary-General 25
United States xv, 11, 15, 37, 43, 50, 56, 62–3, 78–9, 105, 112–3, 119–20, 128, 146–7, 149, 151, 157–8, 165, 179, 187, 189–90, 201
  Constitution 35
  ecological footprint 78, 104
  EPI index 165
  HPI index 158
UNCTAD 83, 147–8
UNDP 6, 80
UNEP 4–6, 38, 148
urban land 26, 29, 69, 183
Urban Eco-Balance tax 68–70, 190–91
urban sprawl 70
Utopia 25
US Congress 56
US dollar 119, 121–2, 125

values 48, 51, 122, 182, 185, 198, 201, 204
Vanhanen, H. 27
variety 44, 47
  in economic systems 53
  strength in 51, 158
Venuzuela 176
Vickers, Tony 59
Victor, P. 16
Voss, P.R. 187

war 161, 201, 205
  small conflicts 76, 183
Washington Consensus 51
waste 60, 95–96, 184
Waste Electrical and Electronic Equipment (WEEE) Directive 166
water boreholes 38
Watkiss, Paul 67
Waxman, Henry 56
WCED 7, 21
wealth 80–81, 104
Weizsäcker, E. 169
Welzel, C. 36
West, the 33, 48–9, 78–9, 132, 138, 187, 201
  Western model 33, 48
white goods 94
wilderness 26, 27
Williamson, J. 52
WIR Bank 123
Woodin, Michael 12
World Bank 6, 14, 18, 33, 38, 48–9, 138, 155, 184, 189, 196
World Business Council for Sustainable Development (WBCSD) 188
world economy 3, 4, 13, 22, 43–4, 52, 54, 64, 83, 134, 138, 147, 153, 172–3, 175, 196, 198, 200
World Forum on Enterprise and the Environment 15, 203
world leaders 37, 53, 159, 203
World Meteorological Organization (WMO) 4
world money 122
World Sustainable Trade Organisation (WSTO) 152

World Trade Organization, *see* WTO
  Committee on Trade and Environment (CTE) 87
world society 9, 37, 44, 155, 157, 167, 172, 196, 202
World Summit on Sustainable Development (WSSD) 23
WTO 4, 38, 87, 89

Yugoslavia 50

Zambia 68
Zimbabwe 29, 68
  HPI index 158